ORGANICS ANALYSIS USING GAS CHROMATOGRAPHY/ MASS SPECTROMETRY—

A Techniques & Procedures Manual

by

WILLIAM L. BUDDE

and

JAMES W. EICHELBERGER

Environmental Monitoring and Support Laboratory
Office of Research and Development
U.S. Environmental Protection Agency
Cincinnati, Ohio

ANN ARBOR SCIENCE
PUBLISHERS INC
P.O. BOX 1425 • ANN ARBOR, MICH. 48106

This manual has been reviewed by the Environmental Monitoring and Support Laboratory-Cincinnati, U.S. Environmental Protection Agency, and approved for publication. Mention of trade names or commercial products does not constitute endorsement or recommendation for use.

Published 1979 by Ann Arbor Science Publishers, Inc.
230 Collingwood, P.O. Box 1425, Ann Arbor, Michigan 48106

Library of Congress Catalog Card Number 79-88484
ISBN 0-250-40318-8

Manufactured in the United States of America

FOREWORD

Environmental measurements are required to determine the quality of ambient waters and the character of waste effluents. The Environmental Monitoring and Support Laboratory-Cincinnati conducts research to:

- Develop and evaluate techniques to measure the presence and concentration of physical, chemical, and radiological pollutants in water, wastewater, bottom sediments, and solid waste.

- Investigate methods for the concentration, recovery, and identification of viruses, bacteria and other microbiological organisms in water; and to determine the responses of aquatic organisms to water quality.

- Develop and operate an Agency-wide quality assurance program to assure standardization and quality control of systems for monitoring water and wastewater.

This publication is the result of an Agency-wide effort to document broad spectrum methods needed for the analysis of organic compounds in environmental samples. Broad spectrum methods are required for situations where the nature of the organic compounds that may be present is unknown. Federal agencies, states, municipalities, universities, private laboratories, and industry should find this manual of assistance in monitoring and controlling organic pollution in the environment.

Dwight G. Ballinger
Director
Environmental Monitoring and
Support Laboratory-Cincinnati

This procedural manual defines the areas of applicability of gas chromatography/mass spectrometry in environmental analysis. The manual includes sample preparation methods specifically adapted to this measurement technique, data processing and interpretation methods, quality control procedures, quantitative analytical techniques, utility software requirements, preventive maintenance recommendations, systematic trouble shooting methods, and a selected bibliography.

TABLE OF CONTENTS

LIST OF FIGURES

LIST OF TABLES

DATA SYSTEM RULES AND CONVENTIONS USED IN THIS MANUAL

1. Each line of response to a computer program entered from a keyboard by a user must be terminated by pressing the return key. This is not indicated in sample dialogue.

2. User responses in sample computer dialogue are underlined.

3. A user response to a computer program of *YES* may be abbreviated *Y*; a user response to a computer program of *NO* may be abbreviated *N*; in most programs an *N* may be replaced by merely pressing the return key. In a few programs a *NO* response will not be accepted and a negative input must be made by pressing the return key. These exceptions are indicated in the sample dialogue.

4. Most user entered commands may be abbreviated with the first four characters.

5. A user may interrupt computer prompts at the keyboard if he understands the correct response before the output is complete.

6. A user may usually return to the SELECT MODE: prompt by holding down the CTRL key while pressing the L key. This is abbreviated CTRL/L frequently.

7. The compound perfluorotributylamine will be abbreviated PFTBA. This is also sometimes known by the 3M company trade designation FC-43.

LIST OF PDP-8 DATA SYSTEM PROGRAMS

William L. Budde is in charge of the Advanced Instrumentation section of the U.S. Environmental Protection Agency's Environmental Monitoring and Support Laboratory in Cincinnati, Ohio. He received a PhD in organic chemistry from the University of Cincinnati in 1963. His principal experience has been in structural organic and analytical organic chemistry with emphasis on the applications of mass spectrometry and nuclear magnetic resonance. He has been employed since 1971 by the U.S. EPA to develop methods for the identification and measurement of organic environmental pollutants using gas chromatography/mass spectrometry. Other research activities include the application of small computers to the automation of laboratory measurements.

Dr. Budde is a member of The American Chemical Society and the American Society for Mass Spectrometry. He has lectured widely and published over thirty scientific journal articles and reports.

James W. Eichelberger is a research chemist at the U.S. Environmental Protection Agency, Environmental Research Laboratory in Cincinnati, Ohio. He received his BS degree in chemistry from Xavier University in 1960 and worked with the U.S. Public Health Service measuring various types and sources of low-level radiation in the environment. In 1964, he began working in gas chromatography for the U.S. Department of the Interior, and was instrumental in the development of gas chromatographic procedures to identify and measure polychlorinated biphenyls, various types of pesticides, and other classes of environmental contaminants. Since 1971, Mr. Eichelberger has been working in computerized gas chromatography/mass spectrometry developing analytical and quality assurance procedures for environmental analyses.

He is a member of the American Society for Mass Spectrometry and has served for the past five years as editor of the *EPA Mass Spectrometer Users' Newsletter*. He is author or coauthor of a number of publications in the areas of trace organics analysis in the environment and gas chromatography/mass spectrometry methodology.

CONTRIBUTORS

Special recognition is required for several individuals who made significant contributions to this manual. These individuals are:

Ann Alford
Environmental Research Laboratory, EPA
Athens, GA 30601

Thomas A. Bellar
Environmental Monitoring and Support Laboratory
Cincinnati, OH 45268

Harvey W. Boyle
National Enforcement Investigation Center, EPA
Denver, CO 80225

Robert D. Kleopfer
Region VII, EPA
Kansas City, KS 66115

D. Craig Shew
Robert S. Kerr Environmental Research Laboratory, EPA
Ada, OK 74820

ACKNOWLEDGEMENT

The authors wish to acknowledge the many others who participated in the development or review of the manual.

Herbert J. Brass
Division of Technical Support
Office of Water Supply, EPA
Cincinnati, OH 45268

Mike Carter
Environmental Research Laboratory, EPA
Athens, GA 30601

Emile Coleman
Health Effects Research Laboratory, EPA
Cincinnati, OH 45268

B. F. Dudenbostel
Region II, EPA
Edison, NJ 08817

Jack D. Henion
New York State College of Veterinary Medicine
Ithaca, NY 14853

Doug Kuehl
Environmental Research Laboratory, EPA
Duluth, MN 55804

Robert Lingg
Health Effects Research Laboratory, EPA
Cincinnati, OH 45268

Bruce G. Logan
Bowne Information Systems
Washington, DC 20036

William Loy
Region IV, EPA
Athens, GA 30601

Robert Melton
Health Effects Research Laboratory, EPA
Cincinnati, OH 45268

William Middleton
Environmental Monitoring and Support Laboratory, EPA
Cincinnati, OH 45268

David J. Munch
Division of Technical Support
Office of Water Supply, EPA
Cincinnati, OH 45268

Curt Norwood
Environmental Research Laboratory, EPA
Narragansett, RI 02882

Frank Onuska
Canada Center for Inland waters
Burlington, Ontario, Canada

James F. Ryan
Health Effects Research Laboratory, EPA
Research Triangle Park, NC 27711

Jack M. Teuschler
Environmental Monitoring and Support Laboratory, EPA
Cincinnati, OH 45268

Richard J. Thompson
Environmental Monitoring and Support Laboratory, EPA
Research Triangle Park, NC 27711

Gilman Veith
Environmental Research Laboratory, EPA
Duluth, MN 55804

Jacqueline Waitts
Bowne Information Systems
Washington, DC 20036

Virgil L. Warren
Region VI, EPA
Houston, TX 77036

Ron Webb
Environmental Research Laboratory, EPA
Athens, GA 30601

Karen Zieverink
SouthWestern Ohio Regional Computer Center
Cincinnati, OH 45221

CHAPTER 1
INTRODUCTION

The purpose of this manual is to organize in one place a variety of information that is needed to identify and measure organic compounds in environmental samples using computerized gas chromatography-mass spectrometry (GC/MS). The application of a mass spectrometer as a universal, yet extremely selective and sensitive detector in gas chromatography has revolutionized the identification and measurement of organic compounds. The efficient use of GC/MS was made possible by the development of the relatively low cost digital minicomputer during the late 1960's. During the subsequent years a large number of books, review articles, and journal articles appeared which describe the basic concepts of the GC/MS instrumentation, and applications to a wide variety of organic analytical problems. A number of these books and articles are cited in the Bibliography in Chapter 10 of this manual. However in the environmental field the specific detailed information that is required to utilize this powerful tool is scattered among a variety of sources, and some vital information is not written down at all. The emphasis of this manual is on the experimental details of the GC/MS methodology.

There are three distinctly different applications where environmental measurements by GC/MS are being or will be used widely. First, there is the broad spectrum (BS) organics analysis which has the goal of seeking a broad spectrum picture of whatever is present in a sample as a major or minor component. This kind of analysis is not guided by a predetermined list of compounds to be measured. The BS approach was made possible by the development of tools such as computerized GC/MS, and this approach is emphasized in this manual. The BS approach is most appropriate for samples that are likely to contain unexpected components.

Secondly, the application of GC/MS appears more cost effective than conventional GC methods for routine monitoring of relatively large numbers of target compounds, e.g., 40–60 or more. Some possible reasons for this are presented later. Regardless of the number of compounds monitored, the application of GC/MS for routine monitoring of target compounds is required for samples that are likely to contain unexpected components that are interferences in conventional methods.

Thirdly, the application of GC/MS in real time selected ion monitoring is widely used in the biomedical and other fields, and is recognized as one of the most accurate, precise, and selective tools known to analytical chemistry. This application requires detailed knowledge of the compound being measured, and its mass spectrum. This application is not emphasized in this manual except for one section in Chapter 6. However it is expected that selected ion monitoring methods will become more important in the environmental field in the future.

There are a number of contrasts between the GC/MS methods in this manual and conventional methods, which are often based on chromatographic techniques, of determining organic compounds. Both types of methods are needed, and each has an appropriate area of application. Conventional methods for organic compounds are detailed chemical-instrumental procedures designed to separate the compound of interest (target compound, TC) from all known interferences, and to measure its concentration with relatively inexpensive, simple, easy-to-use instrumentation. The separation and isolation procedure is often called "clean-up" and is, in effect, a large part of the qualitative analysis. The gas chromatography-electron capture detector procedures for chlorinated hydrocarbon pesticides are examples of this approach. The clean-up for conventional detectors is often quite rigorous and complex to assure that all potential interferences are eliminated. Similarly, because conventional methods often rely heavily on retention indices for the identification of compounds, rigorous control of operational conditions such as flow rates and temperature is required. Quality control measurements, such as retention index standards, must be made at relatively frequent intervals to assure that conditions are well controlled.

By contrast, the GC/MS sample preparation methods described in Chapter 3 were designed to be as simple as possible. Rigorous clean-up is often not required because the mass spectrometer output provides adequate information, in most cases, to reliably identify the compound. Rigorous clean-up is, of course, counter productive in the BS approach. The GC/MS methods do not depend on frequent measurements of standards for determination of retention indices.

Another contrast is that conventional methods usually require a variety of detectors to permit measurements of a broad range of compound types. This is because conventional detectors are designed for some selectivity to further minimize the effects of interfering substances. The mass spectrometer is a universal detector which precludes interferences by the very nature of its output. For monitoring a wide range of compound types, it appears that the additional cost of computerized GC/MS instrumentation may be offset by the need to acquire and maintain only one detector, and the use of simpler, less costly sample preparation and quality control procedures.

Several chapters of the manual (Chapters 2,4,7,8, and 9) describe in detail the operation and maintenance of the Finnigan Corporation's models 1015 and 3000 series GC/MS systems with datasystems based on a Digital Equipment Corporation model PDP-8 minicomputer. This information was included and emphasized because the Environmental Protection Agency (EPA) owned more than thirty Finnigan 1015 and 3000 series GC/MS systems and a similar number of PDP-8 GC/MS datasystems. These models of spectrometers and datasystems are in widespread use, and the general user community should find these specialized chapters of interest and value. Users of other types of GC/MS systems will find information of general interest in section 2.6, Quality Control; Chapter 3, GC/MS Sample Preparations; Chapter 5, Compound Identification; Chapter 6, Advanced Analytical Techniques; Chapter 8, Preventive Maintenance; and, Chapter 10, the Selected Bibliography. Also it should be recognized that other GC/MS systems must support certain concepts described in the more specialized chapters, and the ideas presented in them should be of some assistance to the users of other types of systems. Descriptions of this type include section 2.4, System Zero Adjustment; section 2.5, Calibration Diagnostic Program; and, the general trouble shooting concepts described in Chapter 9.

Figure 1.1 shows a flow chart for the processing of a sample with the BS approach. Several blocks in Figure 1.1 are identified with specific chapters in this manual. In the BS approach, the conventional broad band GC detector, e.g., flame ionization, may be an important screening tool for the busy GC/MS laboratory. Alternatively, if the current sample load is not high, the GC/MS may itself perform the initial screening. The GC/MS Operations and Quality Assurance Chapter (Chapter 2) is identified with the block in Figure 1.1 labeled integer accuracy GC/MS. The importance of reagent blanks and spectra interpretation is discussed in Chapter 5. Chapter 6, Advanced Analytical Techniques, includes information on methods that may be used to obtain more information when an identification cannot be made with integer accuracy GC/MS. The Advanced Analytical Techniques chapter also includes information about quantitative measurements with GC/MS.

Contributions to and reviews of the draft chapters of this manual were made by the participants in the EPA mass spectrometer users group between January, 1975 and October, 1978. This users group was organized by the authors in the spring of 1972 to promote the exchange of technical information among EPA laboratories using computerized GC/MS to identify and measure organic environmental pollutants.

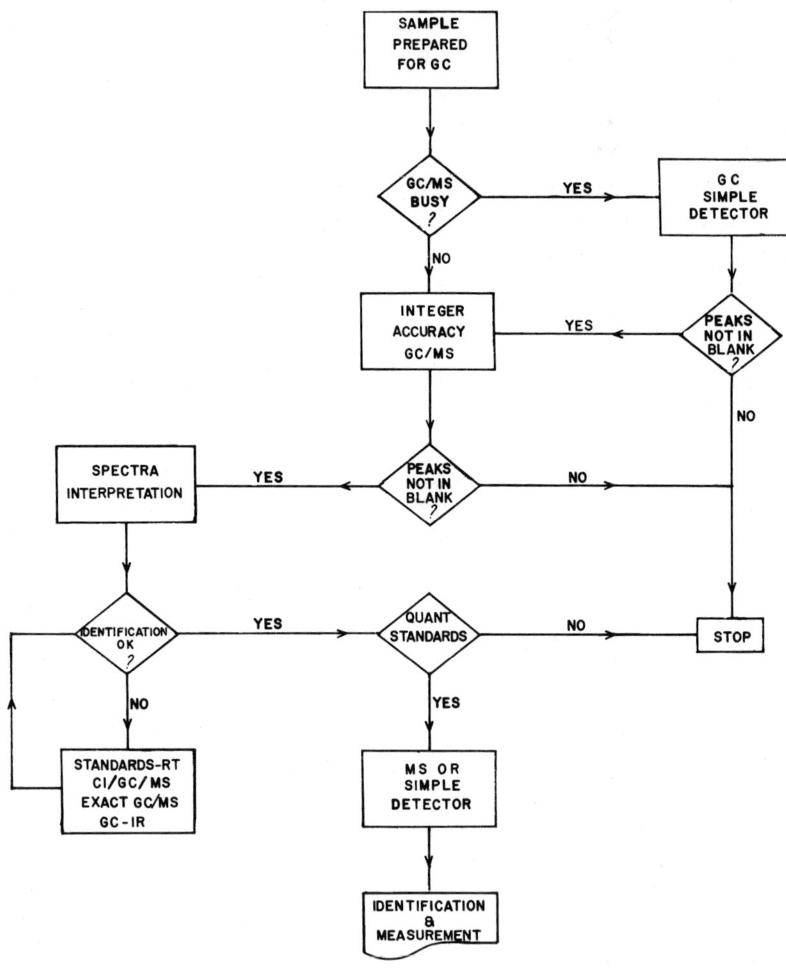

Figure 1.1 Flow of trace organics analysis in a laboratory with GC/MS capability.

CHAPTER 2
GC/MS OPERATIONS AND QUALITY CONTROL

The purpose of this chapter is to provide the basic information required for the successful operation of the computerized gas chromatograph/mass spectrometer (GC/MS). The chapter is oriented to the Finnigan Corporation's models 1015 and 3000 series quadrupole mass spectrometers with data systems that utilize Digital Equipment Corporation model PDP-8 computers. However, certain portions of this chapter, such as the quality control procedure in Section 2.6, are readily applied to other mass spectrometer systems. The quality control procedure is of particular importance because one can be reasonably assured of obtaining good quality mass spectra when the performance criteria outlined in that section are attained.

For initial start-up or whenever the vacuum system is at atmospheric pressure (after changing a separator, cleaning rods, etc.) one should follow the procedures beginning with Section 2.1 and continue through resolution and sensitivity adjustments, data system start-up, system zero adjustments, mass scale calibration, quality control, and sample analysis. However for normal day-to-day operation one would begin with data system start-up followed by system zero adjustment, mass scale calibration, and the quality control tests. Resolution adjustments are usually required only occasionally as indicated by the quality control evaluation. Some suggested operating parameters are given in Section 2.7. These parameters should be considered as guidelines and may not represent optimum operating conditions for every sample which may be encountered in an environmental analysis laboratory.

Recommended shut-down procedures for both short-term (overnight, weekends) and long-term situations are given. Proper shut-down procedures are important to reduce or prevent deterioration or damage to certain expensive components of the mass spectrometer.

2.1 MASS SPECTROMETER START-UP

When the vacuum system is at atmospheric pressure (after changing a separator, cleaning rods, etc.) and the system is completely shut down, the following procedure is recommended for restarting the system.

1. Apply power with the vacuum controller circuit breaker. This supplies line power to the mass spectrometer, the gas chromatograph, and the vacuum system.
2. Make sure that the system to be evacuated is closed to the atmosphere (the GC end of the separator must be capped and the direct inlet system closed).
3. Turn on the forepump (analyzer and batch inlet if present) switch and separator pump switch. These switches activate the mechanical vacuum pumps used to obtain a "rough" vacuum.
4. Check pressures on the forepump pressure meter. This meter monitors the pressure down to 0.01 torr at each forepump. If no leaks are present, readings below 0.1 torr will be obtained within one minute.
5. After pressures of 0.1 torr or below are obtained, turn the vacuum controller function switch to the override position. This turns on the diffusion pump heaters. Since the override position bypasses the protection circuits, the mass spectrometer should not be left unattended during this time.
6. After waiting approximately 45 minutes for the diffusion pumps to heat, attempt to light the Bayard-Alpert pressure gauge by depressing the start button on the vacuum controller. Depressing this button for 1–2 seconds will turn on the ion gauge filament when the system pressure is below 10^{-3} torr. If the filament fails to light, wait five minutes and repeat. The life of the ion gauge filament is shortened by high pressure operation. If the filament stays on at 10^{-4} torr or greater, turn the function switch to off, then back to the override position and wait for lower pressure to be obtained.

   The zero control on the vacuum controller module permits electrical zeroing of the high vacuum meter. This can be done while the ion gauge filament is off and the meter switch is in the override position. Adjust the zero control until the meter reads 1×10^{-7} torr.
7. After a stable vacuum operation of 10^{-5} torr or lower is achieved, move the vacuum controller function switch from override to protect. Normally several hours are required to obtain pressures of

10^{-7} torr or lower. The mass spectrometer can be safely utilized, however, with an analyzer vacuum of 10^{-5} torr or less.

8. Adjust temperatures of the analyzer manifold, transfer line, batch inlet, and interface as desired with the heater controls on the vacuum controller. These temperatures should be monitored occasionally with an external pyrometer of known accuracy. In general, the analyzer manifold should not be heated, and the separator (interface) termperature should be a few degrees higher than the maximum GC column temperature to be utilized. A temperature of about 250C is adequate for the transfer line. The batch inlet temperature should be sufficiently high to permit appreciable vaporization of any liquid to be introduced via the batch inlet (generally 100–200C).
9. Turn on the line power switch on RF/DC power supply (1015 only).
10. Turn power on for high voltage supply to electron multiplier (1015 only).

2.2 RESOLUTION AND SENSITIVITY ADJUSTMENTS

Resolution and sensitivity adjustments are necessary with a quadrupole mass spectrometer after the system has been shut down for cleaning or replacing the ionizer, mass filter, or the electron multiplier. They are also necessary for other reasons such as after changing a component on a printed circuit (PC) board in a power supply or in the DC/RF generator. Before the mass spectrometer resolution and sensitivity can be adjusted, the electronic components must be balanced. The DC zero and balance procedure that follows was written for the Finnigan model 1015. For the Finnigan 3000 series, the corresponding procedure is described as a ROD DC BALANCE ADJUSTMENT on page 1–16 of the 3000 series systems maintenance manual. Figure 2.1 is a functional schematic of the components used in the model 1015 tune-up.

DC ZERO AND BALANCE

1. Set the mass spectrometer as follows:
   RF/DC Power Supply (bottom panel):
   Line Power - On
   Standby/Operate - Operate
   RF/DC Generator:
   Ionizer: Off
   Electron Multiplier Voltage: Off

Scan Time: 0.1 sec
Scan Time Vernier: Fully clockwise
Scan Mode: Repetitive
First Mass Selector: 000
Last Mass Selector: 000
Scope Sensitivity: 0.5V/CM

2. Set the normal/test switch on the rear of the oscilloscope to "test" position.
3. Zero the oscilloscope by adjusting the trace to the center of the screen. Keep it zeroed during the following procedure by momentarily disconnecting the oscilloscope probe from the test point being observed and readjusting the trace.
4. Set the mass range selector on the model 1015 RF/DC generator to a detent position. This removes the tuning capacitor from the RF generator thereby curtailing the output of RF and places PC-5 circuitry under the mid and high mass mode of control. The mass selector is a five position switch having one dentent position between the high and medium range and another between the low and medium range.
5. Turn R-81 on PC-5 (behind the removable panel on the RF/DC generator) fully counterclockwise then turn clockwise 10 full turns. This places R-81 approximately in the center of its adjustment range when amplifier A4 is balanced.
6. Connect the oscilloscope test probe (x10) to TP-3 on PC-5.
7. Turn R-185 fully clockwise until a clicking sound is observed; then turn it counterclockwise until the oscilloscope trace shifts downward about 1.5 cm (15 volts). This balances amplifier A4 and allows R-81 to have the maximum range of adjustment. Connect the oscilloscope test probe to TP-4.
8. Turn R-189 clockwise; then turn it counterclockwise until the oscilloscope display shifts upward about 1.5 cm (15 volts). This balances amplifier A5 (see schematic of PC-5). The 15 volt jumps observed on the oscilloscope are the saturation voltages of amplifiers A4 and A5. R-185 and R-189 influence the range of R-81. If one lacks sufficient range while adjusting the low mass resolution, R-185 and/or R-189 may not be adjusted properly.
9. Switch the mass range selector to 50–750.
10. Set first mass selector to 000 and last mass selector to 750.
11. Connect one end of the rod balance network to TP-5 and the other end to TP-6 on PC-5 (rod balance network = two 100K, 1% resistors connected in parallel to a test point).
12. Connect test probe to the junction between the two resistors.

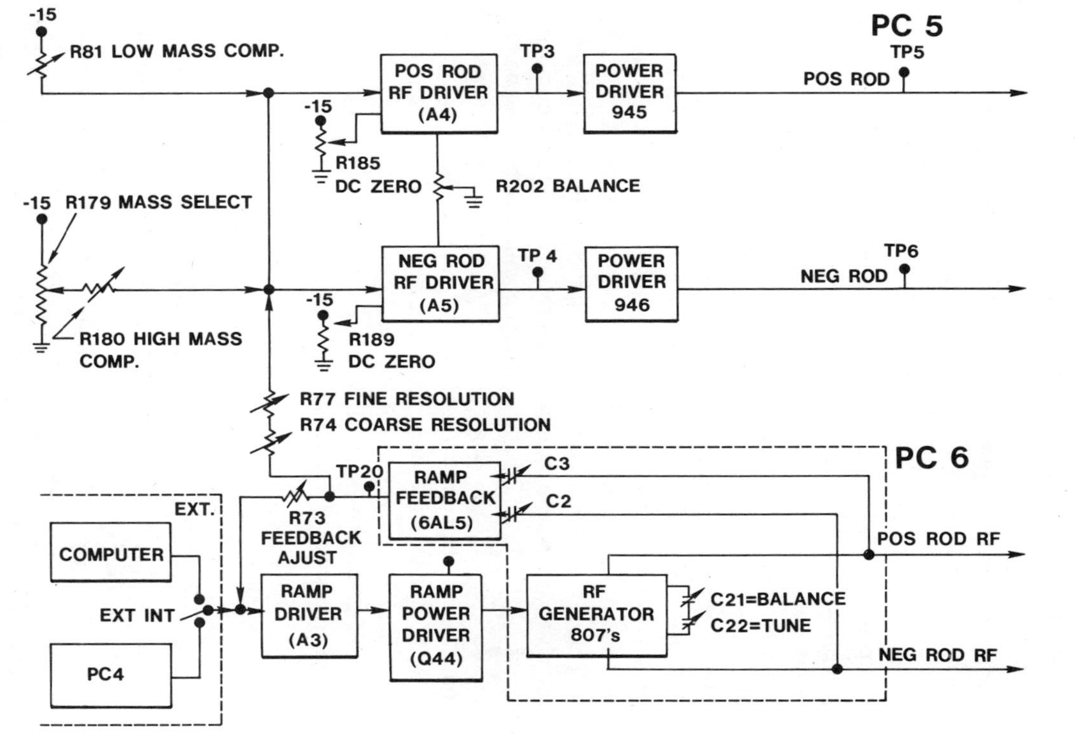

Figure 2.1 Rough functional schematic showing components used in the tune-up of a Finnigan model 1015 in the high mass range.

13. Set oscilloscope sensitivity to 20 millivolts/cm.
14. Adjust R-202 until the trace on the scope is horizontal; then adjust R-189 until the trace is at 0 volts. If R-189 needs to be changed more than 2 turns, go back to R-202 and repeat adjustment.
15. Set the normal/test switch on rear of oscilloscope to the normal position.

After the electronic components have been balanced, one should proceed with the resolution adjustment for 50–750 mass range.

TUNE-UP WITH PERFLUOROTRIBUTYLAMINE

The tune-up procedure that follows was written specifically for the Finnigan model 1015 and 3000 series spectrometers. The same basic philosophy also applies to the model 4000 spectrometer, but the detailed procedure is somewhat different and not described in this manual. Potentiometer (pot) designations (e.g., R-81) refer to the model 1015 only. These pot reference numbers are marked on PC-5 which is behind the removable panel of the RF/DC generator. In parentheses after the 1015 pot number is the pot function name. In the Finnigan 3000 series spectrometers these names are marked on the RF/DC controller board in the RF/DC generator. However, the 3000 series pot numbers, which are referenced in the manuals and on schematics, do not correspond to the pot numbers used in the model 1015. Table 2.1 shows function names and corresponding pot designations for the model 1015, 3000 series, and model 4000 spectrometers. Similar considerations apply to test points.

The EPA tune-up procedure is essentially the same as that given in Finnigan manuals except that certain modifications are made to facilitate generating spectra similar to that obtained with magnetic deflection instruments. These modifications are as follows:

1. There is no optimization of signal intensity or resolution at masses 16–18. This spectral region is rarely, if ever, important in organic analysis.
2. The 10% valley definition of resolution is employed for tune-up at masses 502–503. This improves relative signal intensity in this region of the spectrum.
3. The magnet position, ion energy, and extractor voltages are adjusted for maximum signal intensity and peak shape at masses 502–503.
4. Guidelines for relative abundance criteria are included to assist in achieving performance consistent with magnetic deflection spectrometers.
5. Electron energy is adjusted for optimum signal intensity across the 50–750 amu range.

Table 2.1. Functions and Corresponding Potentiometer Designations for Finnigan Mass Spectrometers

Function	Model 1015	Models 3000-3300	Model 4000
Low mass compensator	R-81	R-35	R-45
Coarse resolution	R-74	R-30	R-65
High mass threshold	R-179	R-29	
High mass resolution	R-180	R-28	R-73
Fine resolution control	R-77	Resolution (front panel)	(front panel)
Balance for amplifier A4	R-185	not applicable	
Balance for amplifier A5	R-189	not applicable	
DC Balance	R-202	R-42	

Resolution adjustment requires the introduction of perfluorotributylamine (PFTBA) into the mass spectrometer. PFTBA has known mass peaks and adjacent ^{13}C peaks up to approximately mass 600. Significant ions are at the following mass numbers: 69—70, 131—132, 219—220, 264—265, 414—415, 502—503, and 614—615.

Using the 50–750 setting on the 1015, one is capable of calibrating and scanning from 10 to 750 amu. Therefore, for normal operations only the high mass range is adjusted.

1. Set the ion source, preamp, oscilloscope, and DC/RF generator controls as follows:

Control	Model 1015	3000 Series
Ionizer (Filament)	On	On
Ionization (Emission) Current	0.5 milliamp	0.5 milliamp
Electron Energy	70 ev	70 ev
Extractor	8v	4v
Ion Energy	5v	2v
Lens	- 100v	40v
Preamp Sensitivity	10^{-6} amps/v	10^{-6} amps/v
Preamp Filter	$>$3000 amu/sec	$>$3000 amu/sec
Scope Sensitivity	0.5 v/cm	0.5 v/cm
Electron Multiplier	Cu/Be, 3000v	Cont. Dyn., 1600v
Mass Range Selector	High	Not Applicable
Scan Time	0.1 sec	GC
Scan Time Vernier	Fully Clockwise	Not Applicable
Record Switch	Off	Not Applicable
Scan (control) Mode	Repetitive	On
First Mass	0	0
Last Mass	750	750

2. Inject 0.5 ul of PFTBA into either the batch inlet, if you have one, or the variable leak inlet system and admit a small amount into the mass spectrometer. Look at scope to observe a mass spectrum. If no spectrum is observed, double check all settings given in step 1 and admit additional PFTBA to increase the system pressure at least a factor of five. If no spectrum is observed, one may continue with step 3, but serious problems are apparent and trouble shooting may be required.
3. Turn R-179 (high mass threshold) fully counterclockwise and R-180 (high mass resolution) fully clockwise.
4. Adjust the first mass and last mass controls so that the peaks at 219 and 220 amu appear on the oscilloscope as in Figure 2.2.

5. Adjust R-77 (resolution-front panel) for the resolution between 219 and 220 as in Figure 2.2. If R-77, the fine adjustment, lacks sufficient range, set it to the middle of its range, and adjust R-74 (coarse resolution) to obtain roughly what is illustrated in Figure 2.2. Then go back to R-77 for the fine adjustment. The resolution observed on the scope is approximate and not necessarily the actual resolution. In general, if resolution on the scope is not quite baseline, the actual resolution will be adequate.
6. Set the first mass and last mass controls so that the peaks at 69 and 70 amu appear on the oscilloscope as in Figure 2.3.
7. Adjust R-81 (low mass compensator) for the resolution as illustrated in Figure 2.3. Trimpots R-77 and R-81 are interactive. Therefore steps 4 thru 7 must be repeated several times before resolution is obtained at both 69–70 and 219–220 amu.
8. Decrease the scope sensitivity until mass 69 is observed completely on the scope. Adjust first mass and last mass controls until mass 69 is on the left side of scope and mass 219 is on right side of scope. Mass 219 should be approximately 35 percent as abundant as mass 69.
9. Set the first mass control so that the peak at mass 264 is on the extreme left side of the scope. Set the last mass control to 750.
10. Turn R-180 (high mass resolution) fully counterclockwise.
11. Slowly turn R-179 (high mass threshold) clockwise until a large peak appears on the right side of the scope. Continue turning R-179 until peak 264 on left side of scope <u>just</u> starts to increase. This allows R-180 to influence masses down to mass 264.
12. Turn R-180 clockwise until the large peak diminishes and changes into spectral lines. This completes a rough tune of R-180.
13. Set first mass and last mass controls to display the masses 502 and 503. Adjust R-180 so that resolution is as in Figure 2.4. This completes a fine tune of R-180.
14. At this point, with peaks 502 and 503 displayed on the scope, adjust the source magnet position, extractor voltage, and ion energy for maximum peak height and best peak shape. In general the collector current should be 70–90% of the emission current.
15. Adjust first mass and last mass controls to display the entire PFTBA spectrum. Then adjust the electron energy for maximum signal intensity. The voltage observed on the meter should then be above 50 and below 80 volts.
16. The mass spectrometer should now be ready to calibrate. There is no advantage to checking resolution with the strip chart recorder.

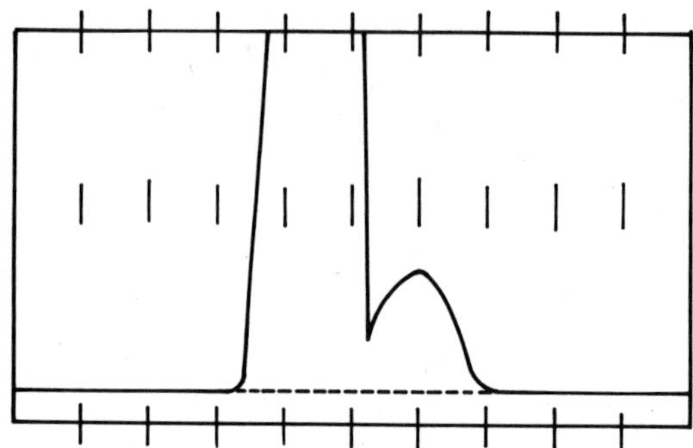

Figure 2.2 Peaks at masses 219 and 220.

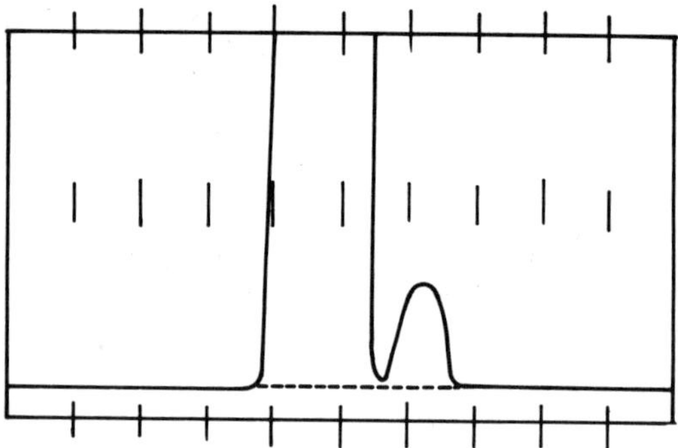

Figure 2.3 Peaks at masses 69 and 70.

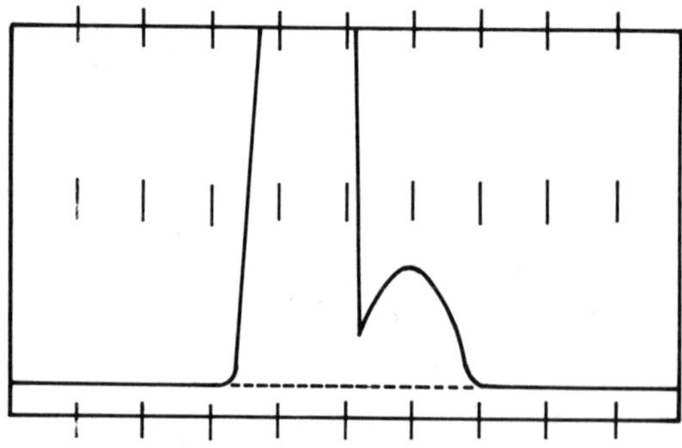

Figure 2.4 Peaks at masses 502 and 503.

2.3 DATA SYSTEM START-UP (PDP-8)

This section describes in detail the start-up of a data system based on a Digital Equipment Corporation model PDP-8 computer. Power is applied to the computer by turning the console key to POWER. The POWER LOCK position also activates the system, but the front panel is disabled and the user cannot see the blinking lights. Peripheral devices (plotter, keyboard terminal, etc.) may require separate activation of power switches at some installations.

The software operating system programs are stored on a real time system disk or on a real time system Dectape on older systems. The disk should be inserted into disk drive 0 (or the tape mounted on unit 1) and the LOAD/RUN switch set to the RUN position. About two minutes are required for the yellow READY light to illuminate, indicating the disk is up to speed and ready to transfer data.

The HALT and SING STEP switches must be up to run a program. The operating software system is started by running the bootstrap loader program. For systems with a hardware bootstrap loader, the loader program is initiated by pressing the SW switch (adjacent to the POWER key) down, then up. If a hardware bootstrap loader is present, the system prompt SELECT MODE: will be displayed on the console printer or cathode ray tube (CRT). If no hardware bootstrap loader is included in the data system, the loader program must be loaded manually. The instructions for loading and running the disk operating system bootstrap loader are in the following section.

LOADING THE DISK OPERATING SYSTEM BOOTSTRAP PROGRAM

The disk bootstrap program is loaded via the computer switch register. The switch register has 12 switches, which are either up or down. A switch in the up position represents a "1" and a switch in the down position represents a"0". The switch register is divided into four fields of three switches each.

Instructions and address locations used to load the disk bootstrap are given in four digit octal numbers to represent the position of the 12 switches. Actual loading procedure, however, requires that the instructions and address locations be keyed in with binary numbers.

The binary equivalents of the octal numbers used to load the disk bootstrap are given in Table 2.2. Load the disk bootstrap in the computer as follows:

Table 2.2. Octal-Binary Equivalents of the Real Time System Disk Bootstrap Program.

Address					Instruction				
Octal	Binary				Octal	Binary			
7737	111	111	011	111	7200	111	010	000	000
7740	111	111	100	000	6502	110	101	000	010
7741	111	111	100	001	6504	110	101	000	100
7742	111	111	100	010	6512	110	101	001	010
7743	111	111	100	011	1356	001	011	101	110
7744	111	111	100	100	6517	110	101	001	111
7745	111	111	100	101	7200	111	010	000	000
7746	111	111	100	110	6514	110	101	001	100
7747	111	111	100	111	6501	110	101	000	001
7750	111	111	101	000	5347	101	011	100	111
7751	111	111	101	001	6511	110	101	001	001
7752	111	111	101	010	5760	101	111	110	000
7753	111	111	101	011	6505	110	101	000	101
7754	111	111	101	100	7402	111	100	000	010
7755	111	111	101	101	5337	101	011	011	111
7756	111	111	101	110	7757	111	111	101	111
7757	111	111	101	111	1400	001	100	000	000
7760	111	111	110	000	0400	000	100	000	000

1. Set the starting address, octal value 7737, in the switch register.
2. Depress the LOAD ADDR key.
3. Set the first instruction (7200) in the switch register.
4. Depress the DEP key.
5. Set the next instruction (6502) in the switch register.
6. Depress the DEP key.
7. Repeat steps 5 and 6 until all instructions have been deposited.
8. Set the computer switch register to 7737.
9. Depress the computer LOAD ADDR key.
10. Place the rotary switch in the MD position.
11. Depress the computer EXAM key.
12. Check that the first instruction, 7200, is displayed in the computer's lower display.
13. Examine each instruction in sequence to ensure that the bootstrap has been loaded correctly. If a location is found to be in error, the address of the location must first be entered using the load address key, before the correct instruction is entered (using the DEP key).
14. Set the switch register to 7737.
15. Depress the computer LOAD ADDR key.
16. Depress the computer CLEAR key, then the CONTINUE.

The first message displayed by the disk operating system is a designation of the revision level of the software, for example, D7 or E0. The revision described in this manual is E0. The next line displayed is called the system prompt. It is the return point for all programs and the user can always return to it by holding down the CONTROL (CTRL) key while pressing the L key on the console keyboard. A program called EXEC and a file named INIT$$ are used here. EXEC contains the system prompt, the CONTROL and IFSS prompts, and some utility functions. INIT$$ contains the system configuation information and constants (see section 7.1).

2.4 SYSTEM ZERO ADJUSTMENT

This adjustment is extremely important and should be checked frequently, at least once each work day. The zero adjustment defines an electronic threshold or baseline for the acquisition of mass spectrometric data. Any signal from the mass spectrometer that is below the threshold will not be observed by the data system. All signals that are above the threshold will be measured. If the threshold is too high, small but real signals will not be observed. This reduced sensitivity will be accompanied by inaccuracies in ion abundance ratios. If the threshold is too low, there will be a decrease in the signal/noise ratio and the mass spectra will contain many 2–5% relative abundance signals that are meaningless.

The zero adjustment is accomplished under computer control by running the zero program. The program is called by responding ZERO to the system prompt:

```
SELECT MODE: ZERO
MANUAL?: N
AUTOMATIC?: N
ZERO ADJUST?: Y
```

The MANUAL prompt is used to maintain or release computer control of the mass spectrometer. A Y response causes the computer to release control of the spectrometer, that is, put it in the manual mode, and the data system will return to the system prompt. Therefore to use the zero program the user must respond to MANUAL with an N. The AUTOMATIC prompt was intended for a future implementation of an automatic zero adjust. This has not been accomplished yet, and the user should respond N. After a Y response to the ZERO ADJUST?, the data system monitors the output of the mass spectrometer at a particular mass. In earlier versions of the software, for example D2, the zero program monitored a mass at the very low end of the mass scale in the 10 - 30 amu range. In later versions, for example D5, this was changed to monitor a mass in the 450 - 500 amu range. The user should make certain that the later version of the zero program is used for the zero adjustment. In the lower mass range there is often significant background from air components and other common gases. If the zero program is run with the ion source and electron multiplier turned on, as is recommended, this background will cause the threshold to be set too high. In the range of 450 - 500 amu there is very little natural background and a better zero adjustment may be accomplished.

The actual threshold adjustment is made, while the zero program is running, by turning the preamplifier zero adjust control until just one light (octal 0001) is observed flashing on and off in the computer accumulator. Zeroing should be performed with the preamplifier switch in the position to be used during data acquisition. The rotary switch on the PDP-8 front panel must be set to the AC position to display the contents of the accumulator in the front panel lights. If no accumulator lights are displayed, the threshold is too high; if more than one accumulator light is displayed, the threshold is too low.

As previously indicated, the zeroing operation should normally be performed with the ionizer and electron multiplier power applied. This should correct for background noise which exists while the mass spectrometer is acquiring data. Many Finnigan GC/MS operators prefer to zero with no power applied to the ionizer and electron multiplier. In theory, if the ionizer is reasonably clean and the electron multiplier is not noisy, the zero setting, with 0001 showing in the accumulator, should be approximately the same with ionizer and power supply on or off. Background noise, however, often affects the preamplifier zero value.

After the attainment of a satisfactory zero, the CTRL/L key is used to return to the system prompt. If a stable zero adjust cannot be achieved, there is a serious hardware problem and the trouble shooting chapter should be consulted.

2.5 MASS SCALE CALIBRATION

The calibration of the mass scale should be checked at the beginning of each day. See section 2.6. If calibration is required the program CALIBR is used for this purpose. Mass scale calibration, requiring approximately 40 seconds, is accomplished over the mass range or over a superset of the mass range selected for spectrum acquisition. The mass scale calibration compound perfluorotributylamine (PFTBA) is injected into the mass spectrometer, and the system automatically calibrates the mass scale using ions from this compound. A file with the calibration data (mass numbers relative to mass set voltages) is automatically created. It should be noted that this procedure does not calibrate ion abundances.

To calibrate, the operator calls the calibration program:

```
SELECT MODE: CONTROL OR IFSS
CALIBRATE?: Y
CALIBRATION FILE NAME: UP TO SIX CHARACTERS
MASS RANGE: 20–700
MS RANGE SETTING?: H
NON STND CAL?: N
DO CALIB DIAGNOSTIC?: Y OR N
```

The calibration program uses the IFSS mode which is explained in Section 2.7. However, it may be called by responding to the system prompt with CONT or IFSS

The calibration file name defines the file which will contain the current calibration data. The file name is used in each data acquisition run to define the mass set voltages to be applied by the data acquisition routine. If the calibration file name entered already exists in the directory, the calibration routine is aborted, an error message is displayed, and the system returns to the system prompt.

The MASS RANGE prompt specifies the range of mass units covered by the calibration file. Only one mass range may be selected for calibration, it must include masses 69 and 100, and the first and last masses must be separated by a hyphen.

The MS RANGE SETTING? prompt refers to a mass range switch on the model 1015 and earlier spectrometers. The user responds to this inquiry with an H, M, L, or P corresponding to the mass range selected on the mass spectrometer electronics console. H, M, L, and P refer, respectively, to the high-mass range (50-750), mid-mass range (10-250), low-mass range (1-100), and peak identifier mass range (1-500). The 3000 series systems have a single mass range and the user responds with an H. It is recommended that 1015 and earlier model users calibrate the high mass range for most general purpose work.

The NON STND CAL? prompt is used for calibration with PFTBA on instruments with an extended mass range (>750 amu) or for calibrations with any alternative mass calibration compound. For calibrations up to 750 amu of extended mass range instruments with PFTBA, the user responds Y. Another prompt requests the name of the overlay file and the correct response is FC1000. A separate overlay file is required for each alternative mass calibration compound. For example, with tris (perfluoroheptyl)triazine as the mass scale calibrant, the overlay filename is M1000. The user should be aware that calibration to 1000 amu requires a change in a system status word as described in section 7.1 For calibrations of instruments with a mass range to 750 amu, the response is N

For the first calibration the user must respond N to the DO CALIB DIAGNOSTIC? prompt. Once the system has been calibrated, and assuming the calibration file was not deleted, the calibration diagnostic feature may be used. After the response to this prompt is entered, the operator must be sure the ionizer is on, and the pressure of PFTBA is adequate for calibration. When calibration is completed, the operating system returns to the system prompt.

A program called MASDEF may be used to shift calibration masses by a constant amount per 100 amu to correct for fractional masses. The program operates on existing calibration files, and the user has several choices of mass shifts. For example, if a shift of +0.06 amu is selected (one of several possibilities), the program sets up a continuous shift of 0.06 amu at 100 amu, 0.09 at 150 amu, 0.13 at 220 amu, etc. Thus one may use a standard PFTBA calibration file and correct for large fractional masses in saturated aliphatic hydrocarbons while using the higher scan speed available with one sample/amu. Negative shifts are also available and have several potential applications.

CALIBRATION DIAGNOSTIC PROGRAM

An affirmative response to the DO CALIB DIAGNOSTIC? prompt calls the program CDIAGN and leads to another prompt:

```
SELECT MODE: CONTROL OR IFSS
CALIBRATE?: Y
CALIBRATION FILE NAME: UP TO SIX CHARACTERS
MASS RANGE: 33–700
MS RANGE SETTING?: H
NON STND CAL?: N
DO CALIB DIAGNOSTIC?: Y
PREV CALIB FILE NAME: UP TO SIX CHARACTERS
```

After input of the name of the previous calibration filename, which still exists on the disk, the program produces a report on the console printer or CRT. A sample report is shown in Table 2.3.

The first row of the report gives the amplitudes (AMP) of the peaks that were used in the current calibration. The mass numbers (AMU) that correspond to these amplitudes are shown in the bottom row. The second row, labeled DEL N (delta N), shows the respective changes in digital-to-analog converter (DAC) values from the previous calibration file. This is perhaps the most significant set of data in the report. Generally when calibration files are compared, and there has been very little drift since the previous calibration, the DEL N values will be no greater than 5-8. If several days have elapsed between calibrations, DEL N values can be expected to approach 10-20 with the Finnigan 1015. DEL N's significantly greater than these values indicate a drifting mass spectrometer or, less likely, a data system problem. The next row of values labeled T, MS is the time, in milliseconds, over which each ion signal was integrated. The amplitude value must be divided by the integration time at each mass to calculate comparable ion abundance values. The row of values labeled N is the DAC value, in octal, that was supplied to generate the mass set voltage for each mass in the current calibration file. To reiterate, the key value is the drift in DAC value of a given mass between calibrations (DEL N).

An affirmative response to the PLOTTER READY question in the Monitor Report will produce the plot shown in Figure 2.5. The diagnostic program uses the table of mass set voltage values that the calibration

Table 2.3. Sample Report from the Calibration Diagnostic Program.

MONITOR REPORT:										
AMP	0	234	93	726	70	244	71	233	89	107
DEL N	0	0	0	−2	−1	−1	−1	0	0	1
T, MS	0	1	16	1	1	1	4	1	8	64
N	1465	2207	4053	5522	10173	12646	15774	22130	42363	63163
AMU	20	28	50	69	100	131	169	219	414	614
PLOTTER READY?										

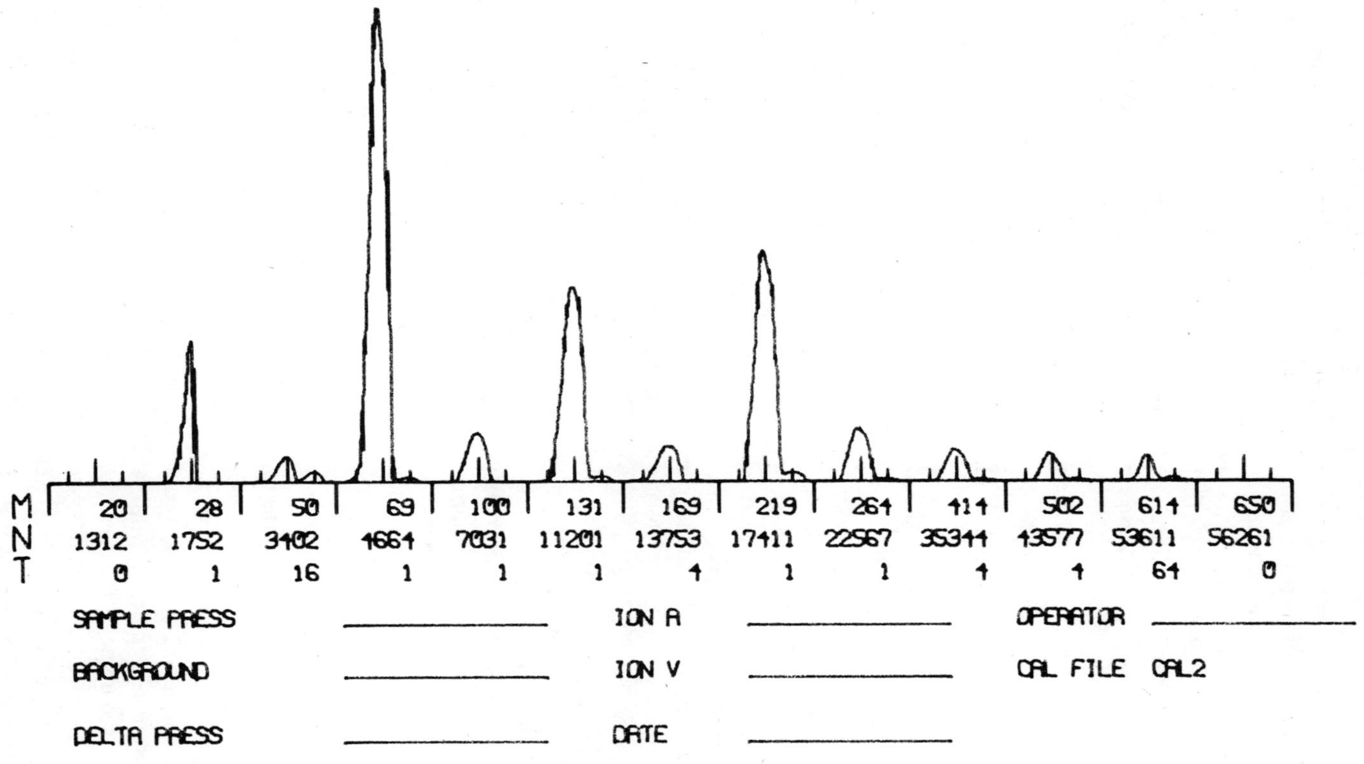

Figure 2.5 Calibration diagnostic program plot

program generated, scans over the ions that were used in the current calibration, and plots the peaks. This plot provides the user with the peak shapes to determine whether resolution was adequate. From these shapes, the user can also determine the condition of the ionizer. Very heavily skewed peaks often indicate that cleaning is required. The plot also gives the operator a record of calibration data on a day-to-day basis. The plot shows relative abundances, the masses of the ions used in calibration, the DAC value settings for each mass, and the integration time spent on each ion. It also allows the user to label the conditions used in calibration that day.

2.6 QUALITY CONTROL

Correct identifications of organic pollutants from gas chromatography mass spectrometry (GC/MS) data require valid mass spectra of the compounds detected. This is independent of the actual method of interpretation of the spectra, i.e., an empirical search for a match within a collection of authentic spectra or an analysis from the principles of organic ion fragmentation. A properly operating and well tuned GC/MS is required to obtain valid mass spectra.

The purpose of the quality control test is to make a quick check - about 15 minutes - of the performance of the total operating system of a computerized GC/MS. Thus with a minimum expenditure of time, an operator can be reasonably sure that the GC column, the enrichment device, the ion source, the ion separating device, the ion detection device, the signal amplifying circuits, the analog to digital converter, the data reduction system, and the data output system are all functioning properly.

An unsuccessful test requires, of course, the examination of the individual subsystems and correction of the faulty component. Environmental data acquired after a successful systems check are, in a real sense, validated and of far more value than unvalidated data. Environmental data acquired after an unsuccessful test may be worthless and may cause erroneous identifications.

It is recommended that the test be applied at the beginning of a work day on which the system will be used and also anytime there is a suspicion of a non-obvious malfunction.

The test was written specifically for the Finnigan quadrupole mass spectrometer equipped with a Digital Equipment Corporation Model PDP-8 computer. However, the test is clearly and readily adapted to any GC/MS system by suitable modification of the detailed procedure. Indeed data from other GC/MS systems was used to establish the abundance criteria of the test (Analytical Chemistry, 47, 995, 1975).

There is a special need to closely monitor the performance of the quadrupole mass spectrometer. Unlike the magnetic deflection spectrometer, the active ion separating device of the quadrupole spectrometer, the rods, is directly contaminated during operation and after prolonged operation is subject to severely degraded performance. Since degraded performance usually affects the high mass region first, the test includes the high mass end criteria. High quality high mass data is important since many environmentally significant compounds have molecular and fragment ions in the 300-500 amu range.

A quadrupole spectrometer which meets the criteria of this test will, in general, generate mass spectra of organic compounds which are very similar, if not identical, to spectra generated by other types of spectrometers. Thus quadrupole mass spectra will be directly comparable to spectra in collections which have been developed over the years with other types of spectrometers.

The reference compound used in this test is available from PCR, Inc., P. O. Box 1778, Gainesville, Florida 32602.

Procedure:

1. Make up a stock solution of decafluorotriphenylphosphine (DFTPP) at 1 milligram/milliliter (1000 ppm) concentration in acetone (or a hydrocarbon solvent). This stock solution was shown to be 97+% stable after 6 months and indications are that it will remain usable for several years. Dilute an aliquot of the stock solution to 10 microgram/milliliter (10 ppm) concentration in acetone. The very small quantity of material present in very dilute solutions is subject to depreciation due to adsorption on the walls of the glass container, reaction with trace impurities in acetone, etc. Therefore this solution may be usable only in the short term, perhaps one week.
2. Adjust the flow from the GC column to 30 cc/min at the exit, attach the column to the spectrometer, and set desired oven temperature. Some suggested GC columns and conditions are listed in Table 2.4. Parameters should be adjusted to permit at least four mass scans during elution of the DFTPP. This will permit selection of a spectrum that is reasonably free of abundance distortions due to rapidly changing sample concentration.
3. Set sensitivity to 10^{-7} and zero instrument using the automatic ZERO program.
4. Set up the real time system for control mode with the following variables: (see section 2.7 for additional information).

```
SELECT MODE: CONT
CALIBRATE?: N
TITLE: DFTPP+ THE DATE
CALIBRATION FILE NAME: CAL
FILE NAME: UP TO SIX CHARACTERS
MASS RANGE: 40–450
INTEGRATION TIME: 8
SAMPLES/AMU: 1
THRESHOLD: press return
RT GC ATTEN: 5
FAST SCAN OPT?: press return
MS RANGE SETTING?: H
MAX RUN TIME: 20
DELAY BETWEEN SCANS (SECS.)?: press return
```

5. Inject 20 ng (2 ul) of the dilute standard into the GC column.
6. After the acetone elutes from the column and is pumped or diverted from the system, turn on the ionizer and start scanning.
7. Note the exact retention time of the DFTPP as it elutes from the column as indicated on the realtime GC. This retention time can be used as a daily check of the condition of your GC column and separator by comparing the values. The retention times should not vary significantly from day to day with identical operating conditions.
8. Terminate the run by pressing CTRL/L, turn the ionizer and multiplier off, and plot the total ion current profile (see chapter 4.)
9. Select a spectrum number on the front side of the GC peak as near the apex as possible and select a background spectrum number immediately preceding the peak.
10. The mass spectrum can be output in various ways including a plot of the full spectrum on the plotter or cathode ray tube or a print of the full spectrum on a printer or cathode ray tube (see chapter 4). A print of a partial spectrum using the following responses is usually convenient:

```
SELECT MODE: OUTP
TOTAL ION CURRENT PROFILE: N
EXTRACTED ION CURRENT PROFILE?: N
PLOT SPECTRUM?: N
PRINT SPECTRUM?: Y
FILE NAME: YOUR FILENAME
SPECTRUM NUMBER.: APPROPRIATE SPECTRUM NUMBER
```

Table 2.4. Suggested GC Columns and Conditions[1]

Dimension (Type)	Packing	Flow Rate	Temp.	R. Time
6′ x 2 mm ID (Glass)	1.95% QF-1 plus 1.5% OV-17 on 80/100 mesh Gas-Chrom Q	30 ml/min	180	4 min.
6′ x 2 mm ID (Glass)	3% OV-1 on 80/100 mesh Chromosorb W	30 ml/min	220	5 min.
6′ x 2 mm ID (Glass)	5% OV-17 on 80/100 mesh Chromosorb W	30 ml/min	220	5 min.
6′ x 2 mm ID (Glass)	1% SP2250 on 100/120 mesh Supelcoport	30 ml/min	170	5 min.

[1]Use any column of your choice which gives at least four mass scans during elution of the DFTPP.

```
PARTITIONED OUTPUT?: Y
MASS RANGE: 51;68-70;127;197-199;275;365;441-443
MINIMUM VALUE %: press return
SUBTRACT BACKGROUND: Y
SPECTRUM NUMBER: APPROPRIATE SPECTRUM NUMBER
BACKGROUND AMPLIFICATION: press return
SAVE SUBTRACTED FILE?: N
NORMALIZE ON: press return
```

The spectrum obtained on the test system should contain ion abundances within limits given for the key ions in Table 2.5. Figure 4.1 is a plot of an acceptable DFTPP spectrum.

If the relative abundances are not within the limits specified, the appropriate adjustments should be made using the proper trimpots on PC-5 of the Finnigan 1015: R-81 (low mass compensator) for the low mass range, R-77 (resolution, front panel) for the medium mass range, and R-180 (high mass resolution) for the high mass range. The magnet position and ion energy may need to be adjusted very slightly to cancel out front end liftoff. When an acceptable spectrum is obtained, it should be recorded for permanent storage.

Table 2.5. Decafluorotriphenylphosphine Key Ions and Ion Abundance Criteria.

Mass	Ion Abundance Criteria
51	30–60% of Mass 198
68	Less than 2% of Mass 69
70	Less than 2% of Mass 69
127	40–60% of Mass 198
197	Less than 1% of Mass 198
198	Base Peak, 100% Relative Abundance
199	5–9% of Mass 198
275	10–30% of Mass 198
365	At least 1% of Mass 198
441	Less than Mass 443
442	Greater than 40% of Mass 198
443	17–23% of Mass 442

2.7 SUGGESTED OPERATING PARAMETERS FOR SAMPLE ANALYSIS

This section deals with the instrumental portion of the analysis after the sample or sample extract has been properly prepared. Details concerning sample preparation are given in Chapter 3. The first part of this section deals with aspects of the gas chromatography itself and includes a list of some recommended columns. The second part describes the operating parameters required for data acquisition.

GAS CHROMATOGRAPHIC CONSIDERATIONS

In order to achieve optimum separations, the selection of the column and gas chromatography operating conditions are of utmost importance. Certainly the degree of separation of components in a complex mixture has a direct bearing on the ability to make unambiquous mass spectral identifications. Publications dealing with selection of the liquid phase and the solid support, along with other operating parameters are referenced in the bibliography (Chapter 10).

Generally, 6 ft. X 1/4 in. O.D. (.08 in. I.D.) packed glass columns with helium carrier gas flowing at 20–30 ml/min are utilized for environmental samples. Comments on open tubular columns are in section 6.3. Silanized glass wool plugs and silanized solid supports of 80/100 mesh are recommended. While using the GC/MS system, one must be alert for the occurrence of leaks or a plugged separator. Normally with helium flowing through the column at 20–30 ml/min, the separator forepump pressure will be 0.1 torr or higher. A pressure lower than this indicates either a leak upstream from the separator or a plugged separator. The normal high vacuum pressure is 10^{-5} to 10^{-6} torr with helium flowing through the separator. A pressure of 10^{-6} torr or lower indicates a plugged separator.

Injections of methylene chloride or chloroform solutions should be avoided on systems without a diverter valve because injections of more than two microliters may cause an excessive increase in pressure in the mass spectrometer which may automatically shut down the system. Acetone and hexane are acceptable solvents to use. Injections should be made with the ionizer off. The elution of the solvent from the column is indicated by a deflection of the high vacuum pressure gauge. After the pressure has returned to normal the ionizer may be turned on and data acquisition can be initiated. However, for systems with a diverter valve, appropriate measures should be taken to divert the solvent from the mass spectrometer. Under these conditions large injections of even troublesome solvents may be quite possible. With hexadecane or tetralin extracts (extraction with high-boiling solvents for determination of volatile components) data acquisition should be started prior to injection and stopped just before elution of the solvent.

Some representative columns for separation of environmental extracts are listed in Table 2.6. Only well conditioned columns should be used and the maximum operating temperatures should never be exceeded.

Certain compounds can be determined by direct aqueous injection into the gas chromatograph. The ionizer is turned on and data collection is initiated immediately after elution of the water. Some representative columns for aqueous injections are listed in Table 2.7.

DATA ACQUISITION

There are two data acquisition modes available to real time system users. The CONTROL mode utilizes a fixed integration time for each mass that is monitored while in the IFSS mode the integration time is varied as a function of the signal strength. The principal file that must be present on the

Table 2.6. Representative GC Columns for Sample Extracts.

Column	Maximum Operating Temperature
1.5% OV-17 + 1.95% QF-1 on Supelcoport	250C
5% OV-17 on Gas-Chrom Q	300C
3% OV-101 on Gas-Chrom Q	300C
4% FFAP on Gas-Chrom Q	200C
3% Dexsil 300 on Chromosorb G 60/80	400C
1% SP2250 on 100/120 Mesh Supelcoport	375C

Table 2.7. Representative GC Columns for Direct Aqueous Injection.

Column Packing	Application
Chromosorb 101 80/100 mesh	General for Low Molecular Weight Compounds up to MW 200. Water elutes before organics
4% FFAP on Chromosorb W 60/80 mesh	Phenols, Acids
0.4% Carbowax 1500 on Carbopack A	Low Molecular Weight Acids
Tenax GC 60/80 mesh	Amines
Carbowax 400 on Porasil C 100/200 mesh	General for Low Molecular Weight Compounds

disk is named CONTRL. This handles data acquisition for the CONTROL and IFSS modes. For real time display on the CRT the files RTCON and CONDIS must be present. For data acquisition and storage of accurate time data with the remote in base (RIB) interface, the file CONDI1 must be present. For data acquisition on a second disk drive the files TWIN and EXMOD2 must be present. None of these files are directly called by the user

As shown in the examples in Section 2.6 and Chapter 3, data acquisition is initiated by answering CONTROL to the system prompt followed by a NO response to the calibrate question. The operator then enters the required information for each of the remaining prompts which are described as follows:

TITLE: The title, which is later printed or plotted on the data outputs, can include up to 64 characters which identify and describe the samples. Such things as sample number, date, and gas chromatographic conditions may be included here.

CALIBRATION FILE NAME: The calibration file name is the file which contains the current calibration data (see Section 2.5 for details on calibration).

FILE NAME: This input, which may be up to six characters in length, identifies the data acquisition file. This file name is used later to output data.
MASS RANGE: This input specifies the atomic mass units which are to be scanned during data acquisition. One can have a single mass range, a set of up to 8 mass ranges, or, up to 8 individual atomic mass units. The mass range inputs must be entered in ascending nonoverlapping order separated by semicolons. The mass range inputs need not be contiguous.

Prior to initiating data acquisition, the control routine checks the operator inputs against the inputs specified for the calibration file. This check ensures that the mass range is a subset of the values specified in the calibration file. If there is a discrepancy in the mass range setting, (see below) or if the user attempts to run outside of the calibration file range, the run will be aborted, and an error message will be printed on the console or CRT.
INTEGRATION TIME: This input specifies the time in milliseconds over which the mass spectrometer monitors each individual mass. A numeric value of 1 to 4095 is entered for each mass or mass range entry. These are separated by semicolons.
SAMPLES/AMU: This parameter can be used to compensate for noninteger masses (for example, 243.2) when necessary. The number of samples to be taken per atomic mass unit must be in the range of 1 to 10. Normally an input of one is used. This provides sampling at integer mass units only. However, a larger number can be utilized in which the data acquisition program proceeds from the integer amu by 0.1 amu steps until the number of samples per amu has been exhausted. The highest intensity point is then recorded as the peak center. A single numeric value defining the number of samples/amu must be entered for each mass range selected. For example, an entry of 1;4;1 coupled with a mass range entry of 40-100; 101; 102-190 would result in sampling at integer mass units for each mass except 101. Mass 101 would be sampled at 101.0, 101.1, 101.2, and 101.3.
THRESHOLD: This input is a single numeric value which applies to the entire mass range being monitored. The number, which represents an integer multiple of .025% of full scale, defines the background noise threshold of the system. The data acquisition software uses this value as the lower limit for the retention of data. A sample point whose ion intensity does not exceed this limit is discarded and reported as zero intensity. The input range is from 0 to 4095 and the default value is 0. This value modifies the zero setting described in Section 2.4, and should be used with caution.
RT ON CRT?: A negative response to this causes the program to put the real time gas chromatogram on the plotter, and even if the remote in base (RIB) interface is present, no real time clock data is saved. An entry of Y or YG initializes the real time clock capability if the system has a RIB

interface. A Y causes display on the CRT of every third mass spectrum if the integration time is 3 msec or longer; if the integration time is less than 3 msec, the program bombs! A YG causes display of the total ion current profile (TICP) on the CRT. Far more important, when the DATA prompt is printed and the user presses the space bar, the real time clock is zeroed and started. The CRT or TTY bell will ring every two seconds; when the user presses RETURN, data acquisition begins. Thus the clock may be started at injection time and data acquisition started subsequently. Alternatively pressing RETURN immediately after the DATA prompt zeros and starts the clock and data acquisition. The real time clock data is saved with the data file, and the MSSOUT software only (Chapter 4) may be used to output a chromatogram with the X axis in spectrum numbers or time in minutes.

RT GC ATTEN: This is a value from 1 to 8 which attenuates the real time output of the total ion current profile recorded on the digital plotter. The default value is 1. For displays on the CRT, the attenuation of the spectrum is accomplished with switch registers 0–5; the attenuation of the TICP is accomplished with switch registers 6–11. The switch registers may be changed during the course of the run, and on both the larger the octal number the more the attenuation of the display.

FAST SCAN OPTION?: This option modifies the overall scan rate of the system, by reducing the settling time (ordinarily 2.3 milliseconds) between successive mass peaks. A response of yes should be made only if this hardware option has been installed.

MS RANGE SETTING: This indicates the range to be used (High, Medium, or Low) during data acquisition with a model 1015. The operator responds with an H, M, or L, depending on the mass range selected on the mass spectrometer electronics console. Note that the calibration file utilized must also correspond to the mass range setting. This was described more completely in Section 2.5. The 3000 series users should respond with an H

MAX RUN TIME: This specifies, in minutes, the duration of data acquisition. A response of zero or pressing the return key results in only one scan being made. The upper limit for this entry is 2047 minutes.

DELAY BETWEEN SCANS (SECS.): An input here allows one to delay between scans from one second to 2047 seconds. No response results in no delay between scans:

SECOND DISK?: When data acquisition on a second disk (D1) is carried out, the current calibration file must be present on the second disk. The file name created on D1 is exactly the same as the designated file name on D0, and the file on D1 starts over at spectrum number 1. The response to this prompt is YES or NO, depending on whether data acquisition should be continued on the second disk if the first disk becomes filled to capacity.

After all of the prompts have been answered, the computer responds with DATA and acquisition begins when the return key is depressed. Data acquisition will end when the previously stipulated maximum run time is reached or it can be halted manually by simultaneously depressing the CONTROL and L keys.

The time required to scan a complete mass spectrum can be determined from the mass range, the integration time, the number of samples/amu, and the settling time which is normally 2.3 milliseconds. For example, data acquisition in the control mode with a mass range of 40–400, one sample/amu, and an integration time of 8 ms would result in scan times of 3.7 seconds: $(400–40) \times (8 + 2.3) = 3708$ ms. Generally scan times of 3-5 seconds are adequate to produce good quality mass spectra. However, shorter scan times may be necessary for early eluting narrow GC peaks. The scan time must be multiplied by the number of samples/amu, and this could increase scan time significantly.

The same operating parameters may be used for data acquisition at a later time because these parameter values are stored with the data file. Reuse of the same parameters requires that the data file be present on the disk. The program used to implement this capability is called PERCAL. The name of an existing data file is entered and PERCAL retrieves the stored operating parameters, e.g., mass range, samples/amu, etc. and prints them on the CRT or TTY; also PERCAL places them in their proper places in core memory for use in the next data acquisition.

```
SELECT MODE:  PERCAL
FILE NAME?:  TEST
TYPE OF RUN:  CONTROL
RUN TITLE:  DEMO

CAL. FILE NAME:  CAL
NB. OF RANGES:  0001
MASS RANGES:  0035–0500
INTEG.TIME(S):  0008
SAMPLE/AMU:  0001
THRESHOLD:  0000
MAX. RUN TIME:  0025M
DL. BET. SCANS:  0000S
RT GC ATTEN:  0003
NB. OF SPECTRA:  0485
NB. OF POINTS:  0000
FILE NAME?:
```

Entry of another file name will cause the program to access another datafile and repeat the process. Entry of CTRL/L causes a return to SELECT MODE, but with the data from the latest PERCAL saved in core. If the response to SELECT MODE is an asterisk, a new data acquisition may begin with minimum dialogue using the same operating parameters.

```
SELECT MODE:     *
FILE NAME:       TURKEY
MAX RUN TIME:    20

DATA
```

Another program called RSTORE is similar, but has more extensive capabilities for storing information about the sample, etc. and works with new and old data files.

INTEGRATION TIME AS A FUNCTION OF SIGNAL STRENGTH (IFSS)

The IFSS mode is a signal optimization mode which is used when the operator wants the system to adjust the Integration time as a Function of the Signal Strength. The use of this option results in a nearly constant signal-to-noise ratio and reduces the chances of saturation of the detector. Figure 2.6 is a flow chart of the IFSS algorithm. The dialogue begins when the operator responds to the computer prompt SELECT MODE by entering IFSS. The other prompts have the same significance as in the control mode except for the following:

MAXIMUM REPEAT COUNT: This input (from 1 to 4096) specifies the maximum number of times to repeat the specified base integration time if the signal level does not reach the specified upper threshold. In effect, this input coupled with the base integration time allows the operator to specify a maximum integration time (maximum repeat count x base integration time = maximum integration time).

BASE INTEGRATION TIME: The operator responds by entering a number from 1 to 4096, which is the time in milliseconds desired for base integration time. Generally a low base integration time (1-4) is desirable in order to avoid saturation of the detector.

REPEAT COUNT BEFORE CHECKING LOWER THRESHOLD: This input (from 1 to 8) specifies the number of times to repeat the base integration time before testing the lower threshold value.

If the signal strength for a given ion has not reached the lower threshold value, then the integration moves on to the next higher mass ion.This, in effect determines the minimum integration time (repeat count before checking lower threshold x base integration time = minimum integration time).

LOWER THRESHOLD: This value determines the point at which the program ceases the sampling of a given ion when a weak signal occurs. The program collects the signal for the minimum integration time, then if the lower threshold has not been attained, integration continues at the next higher mass. A value from 1 to 8 must be entered. These values correspond to the strength of the signal required to continue integration at a given mass (1 = 0% full scale, 2 = 0.025%, 3 = 0.05%, 4 = 0.10%, 5 = 0.20%, 6 = 0.39%, 7 = 0.78%, 8 = 1.56%). An entry of 1 would mean that integration would continue for the maximum integration time or until the upper threshold is attained.

UPPER THRESHOLD: This value determines the point at which the sampling of a given ion ceases when a strong signal occurs. The program collects the signal until the upper threshold limit is attained or the maximum integration time is reached (whichever occurs first), then continues to the next higher mass. The values (1 to 8) correspond to the signal strength required for integration to stop (1 = 50% full scale, 2 = 25%, 3 = 12.5%, 4 = 6.25%,5 = 3.12%, 6 = 1.56%, 7 = 0.7%, 8 = 0.39%). An entry of 1 would mean that, for those signals where the lower threshold has been attained, integration would continue until the signal was 50% of full scale or until maximum integration time is reached.

Numerous specific applications of these operating parameters are contained in Chapter 3 with the appropriate sample preparation methods. The PERCAL program described above also applies to IFSS parameters.

2.8 MASS SPECTROMETER SHUT-DOWN

SHORT TERM

When the system is to be shut down for short periods of time (overnight, weekends, etc.) the following procedure is recommended:

1. Turn ionizer off. Turn electron multiplier voltage power supply to off or standby. Turn RF/DC power supply to standby (1015).
2. Depress HALT switch on computer. Set the LOAD-RUN switch to LOAD and wait for the white LOAD light to signal that the disk has stopped spinning. Shut down computer terminal and other data output devices. Turn computer power key to OFF.
3. Remove beam from oscilloscope with vertical control or on/off switch.

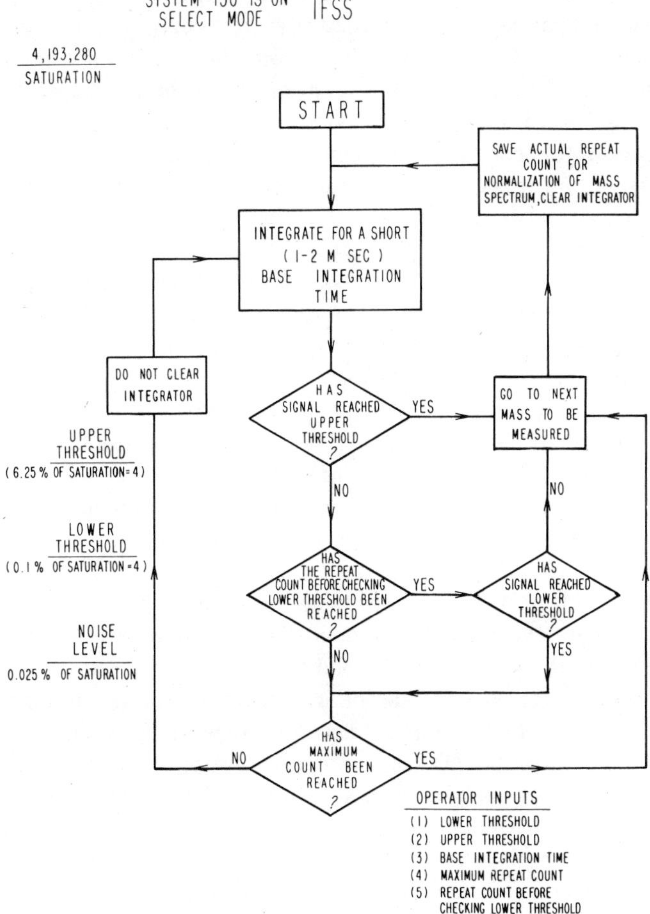

Figure 2.6 The IFSS algorithm flow chart

4. If a diverter valve is present, direct the carrier gas from the mass spectrometer. If no diverter valve is present, one option is to disconnect the column and cap the GC inlet. This allows the spectrometer vacuum system to pump out residual background under a high vacuum. A fritted disk filter in the bulkhead union provides adequate protection from separator clogging during

column disconnect. Alternatively, one may reduce the flow through the GC column to about 10ml/minute, but not disconnect it from the separator. This provides some protection against oil backstreaming into the analyzer in the event of a power failure. Leave the column oven temperature at least 50C below the maximum operating temperature to minimize column bleed.

LONG TERM

When the system is to be shut down for long periods of time or for maintenance purposes (changing separator, cleaning rods, etc.) the following procedure is recommended:

1. Proceed as for short term shutdown.
2. Turn RF/DC line power switch to off (1015).
3. Remove GC column and cap the separator port. Cool column and turn off helium flow through the GC.
4. Turn power off/protect/protection override switch on the vacuum controller to off. This turns off the diffusion pumps, but the mechanical pumps will remain on.
5. Wait at least 45 minutes for the diffusion pumps to cool. Check to make sure that the diffusion pump base is cool to the touch.
6. To vent the system when a solid inlet is present, insert a hose from the regulator on a cylinder of dry helium or nitrogen into the solid inlet assembly. Set no more than 5 psi pressure from the regulator. Tighten the retaining knob on valve assembly. Open both the high vacuum and analyzer valves simultaneously (slowly). When the valves are partially open (you will be able to hear the mechanical pumps pumping the venting gas), throw the main circuit breaker on the vacuum controller to shut off the mechanical pumps. Open the valves all the way, allow the system to fill with helium, and then remove the gas line and close the solid inlet valve.

Exposure of the copper-beryllium electron multiplier to air for extended periods of time (15 minutes or longer) will result in rapid loss of multiplier gain. The multiplier may be damaged permanently. Continuous dynode electron multipliers are not affected by exposure to air. It is recommended that Cu-Be electron multipliers be stored in an inert gas atmosphere or a vacuum desiccator.

CHAPTER 3
GC/MS SAMPLE PREPARATIONS

The sample preparation methods in this chapter emphasize the broad spectrum approach explained in chapter 1. There are no methods that are optimized for a specific compound (parameter) or a closely related group of compounds. Each method is a generalized module with a scope and limitations. Several different modules must be applied to cover the broad range of several hundred thousand compounds that could contribute to the pollutants in any specific sample. In some cases there is significant overlap in the scope of the modular methods. The appropriate choice of a specific module will depend on the objectives of the survey and the scope and limitations of the method.

Clearly there are deficiencies in some of the methods and some environmentally significant compounds may never be amenable to analysis by any method in this chapter. For these classes of compounds, major innovations are required such as the development of liquid chromatography methods. For the present these problems are beyond the scope of this manual.

Many of the methods in this chapter are flexible in that they can be combined or shortened, altered at various stages to fit specific needs, or expanded to cover unanticipated contingencies. Innovation in using these procedures is encouraged insofar that it will improve or otherwise make a particular analysis more efficient. Hopefully, these procedures, in one form or another, can be used for the majority of analyses that confront the GC/MS laboratory.

All of the methods were designed for compatibility with existing GC/MS systems. This is particularly noticeable in the choice of solvents, since some solvents cause excessive pressures in some mass spectrometers and can cause an automatic shut down of the high vacuum system. Generally all methods are designed to be as simple as possible. The reliable qualitative information inherent in the GC/MS method precludes the need for a great deal of clean-up in most samples. Simplicity in methods minimizes the chances of contamination from reagents and glassware. However where appropriate, as in fatty tissue and sediment analyses, fractionation procedures are included in the methods.

Recovery data on spiked samples is presented with many of the methods. It should be recognized that these values reflect the best effort of a skilled and experienced laboratory scientist in one or two laboratories. These data do not guarantee that similar values will be obtained in other laboratories. Recoveries of laboratory control standards and spikes must be determined in each laboratory as part of the quality assurance protocol of the method if the data are used for concentration measurements. Section 6.2 contains a great deal of information on quantitative measurements and quality control.

All of the methods in this chapter were contributed by EPA laboratories. Many were developed or evolved because of the need to have methods oriented to GC/MS that can be used for a particular type of sample or group of compounds.

The majority of the methods are for organics in water, but that is not to imply that analyses of organics in air, sediment, and fatty tissues are less important. In reality, there are fundamentally no differences in the GC/MS aspects of the analyses of environmental samples from different media. The principal differences are in the sampling techniques and preliminary separation procedures. Therefore much of the information found in the section on water samples is applicable to samples from other media. References to these sections are included in the sections on air, sediment, and fatty tissue samples.

In the interest of consistency, each method is described in a similar format. The first part of the method contains a brief definition of the method, advantages and disadvantages, scope and limitations, the rationale for the choices of materials and reagents, detection limits, recovery efficiencies, and any special problems with the method. All detection limits are expressed in terms of a 33-450 amu mass spectrum with average noise $<1\%$ of the base peak. A reagents section is included if reagents are required with emphasis on the purification of materials. An equipment section lists and describes unusual or new equipment only. There is no listing of standard, conventional glassware and other equipment that should be found in any well equipped laboratory. The procedure section is a step-by-step account of the method itself including unusual sampling and storage considerations, cleaning of glassware and sample containers, requirements for blanks and spikes, GC column selection, temperature programming and other GC conditions, and sample computer dialogue for the GC/MS data system. The suggested GC/MS parameters correspond to the continuous repetitive measurement of spectra and not selected ion monitoring. Complete illustrated definitions of these terms are in section 6.1.

3.1 WATER SAMPLES

Water samples encompass a wide variety of types including processed water for human consumption, lake water, river water, ground water, leachate, municipal wastes, industrial wastes, and treated wastes. Each of the sample preparation methods is applicable to one or more of the water sample types and many references to these applications appear in the following sections.

DIRECT AQUEOUS INJECTION INCLUDING CONCENTRATION TECHNIQUES

Direct aqueous injection is the introduction of a few microliters of water sample into the GC/MS system. This procedure allows the identification of compounds that are not amenable to other methods because they are too water soluble.

The minimum detectable concentration using direct aqueous injection is 1 to 5 mg/1 (ppm) depending on the compound observed. The method is suitable for analysis of many types of samples, expecially industrial and municipal wastes.

The following types of compounds can be detected by GC/MS using the direct aqueous injection procedure: aliphatic hydrocarbons C1 - C7; aromatic hydrocarbons-up to short chain alkylated benzenes; the more volatile alcohols, glycols, aldehydes, ketones, ethers, fatty acids, esters, amines, amides, and halogenated compounds.

Simple distillation can be used to concentrate water samples for direct aqueous injection. This technique has been used to concentrate alcohols, ketones, amines, nitiles, dioxolanes, and pyridines. Chlorinated solvents such as methylene chloride and chloroform also have been identified in distillates of industrial effluents. Table 3.1 shows examples of recoveries of known amounts of several compounds from a waste sample after simple distillation. This previously unpublished data were provided by William Loy of EPA. Approximately 10% of the spiked solution was distilled in these experiments.

The minimum detectable quantity is lowered by distillation to about 1-10 ug/1 in the original sample. It may be necessary to concentrate several liters of water for situations such as the analysis of drinking water where taste and odor problems have been observed. This can be done by using a large distillation flask or by distilling several portions of the sample and compositing the distillates for a final distillation in a semi-micro apparatus. This allows a detection limit lower than 1 ug/1 for some compounds.

Table 3.1. Recoveries of Spiked Organics from Water After a Simple Distillation.[1]

Compound Added	Number of Replicates	Average % Recovery	Range
methanol	4	31	18–66
acetonitrile	4	44	38–54
propionaldehyde	4	70	59–88
methyl ethyl ketone	4	69	62–73
1,4–dioxane	4	23	21–25
n–butanol	6	56	52–61
aniline	5	12	9–21
isopropanol	4	54	52–57
nitrobenzene[2]	1	57	——
o–toluidine[2]	1	20	——
di–n–propylamine[2]	1	60	——
di–n–butylamine[2]	1	68	——

[1]Five hundred ml of boiled distilled waste was spiked at 0.2 mg/1 with the indicated organic chemicals. [2]The spiked water was made strongly basic.

A simple evaporation technique is useful for concentrating organic compounds whose boiling points are too high for steam distillation. This has been applied to the identification of ethylene glycol in water.

No special equipment is required for direct aqueous injection. The distillation concentration technique requires a conventional, all-glass laboratory still with a 1-3 liter capacity.

Install a conditioned column in the gas chromatograph. Special care should be taken to flush the column sufficiently with helium to remove any residual air from the packing before heating. Be sure the column was conditioned to minimize bleed during the sample run. Columns which are applicable to direct aqueous analysis are listed in Table 2.7. Temperature programming from 70C to 220C at 8 degrees per minute may be utilized. Inject 2-8 microliters of the water sample with the ionizer off. With the pressure about 10^{-5} torr, turn on the ionizer, start data acquisition, and begin temperature programming. Halt data acquisition after about 35 minutes.

Either the control mode or the IFSS mode may be used for data acquisition. The sample dialogue below contains some suggested parameters.

```
SELECT MODE: CONTROL
CALIBRATE?: N
TITLE: UP TO 64 CHARACTERS
CALIBRATION FILE NAME: CURRENT FILE
FILE NAME: UP TO SIX CHARACTERS
MASS RANGE: 20–260
INTEGRATION TIME: 17
SAMPLES/AMU: 1
THRESHOLD: press return
RT ON CRT?: N
RT GC ATTEN: 1 - 8
FAST SCAN OPT?: N
MS RANGE SETTING?: H
MAX RUN TIME: 30
DELAY BETWEEN SCANS (SECS.)?: press return
```

```
SELECT MODE: IFSS
CALIBRATE?: N
TITLE: UP TO 64 CHARACTERS
MASS RANGE: 20–260
SAMPLES/AMU: 1
MAX RPT COUNT: 32
BASE INTEGRATION TIME: 1
RPT COUNT BEFORE CHECKING LOWER THRESHOLD: 8
LOWER THRESHOLD: 4
UPPER THRESHOLD: 4
RT ON CRT?: N
RT GC ATTEN: 1 - 8
FAST SCAN OPT?: N
MS RANGE SETTING?: H
MAX RUN TIME: 30
DELAY BETWEEN SCANS (SECS.)?: press return
```

If no peaks are observed, a concentration procedure may be applied. For distillation of industrial effluents, transfer 500 ml of sample to a one liter distillation flask, add boiling stones, and collect 25–30 ml of prime distillate. This prime distillate may be redistilled in a semi-micro distillation apparatus with the collection of one ml of secondary distillate. The prime or secondary distillate is analyzed by direct aqueous injection as above.

For distillation of samples that are suspected to contain less than one ug/1 of an organic compound, collect only 10-15 ml during the first distillation. Discard the pot residue and add an additional two liters of sample. Distill as above and composite the prime distillates by collecting the second 10-15 ml in the same collection flask.

A simple evaporation may be accomplished in a beaker or evaporating dish on a hot plate. The residue in this case is analyzed as above.

INERT GAS PURGING AND TRAPPING

Gas purging and trapping is a method for the isolation, concentration, and determination of low boiling organics in water. The method uses finely divided gas bubbles passing through the water sample to transfer organic compounds from the aqueous to the gas phase. The compounds concentrate by adsorption on a porous polymer trap at room temperature as the purge gas is vented. The compounds are subsequently desorbed at elevated temperature by backflushing with a carrier gas into the gas chromatographic system. The method may provide both qualitative and quantitative information. Purging may be accomplished at ambient or elevated temperatures with helium or another inert gas.

Equipment. The equipment required for this method consists of a purging device, a trap, and a trap heater or desorber. Figure 3.1 shows construction details for an all glass purging device with a 5 ml sample capacity. The glass frit at the base of the sample volume allows the finely divided gas bubbles to pass through the sample while the sample is restrained above the frit. Gaseous volumes above the sample are kept to a minimum to eliminate dead volume effects, yet sufficient space is allowed to permit most foams to disperse. The inlet and exit ports are constructed of heavy walled quarter inch glass tubing to permit leak free removable connections with finger tight compression fittings containing Teflon ferrules. The removeable foam trap is optional and recommended for samples that foam. A 25 ml capacity purging device is recommended for use with a mass spectrometer GC detector.

Figure 3.2 shows a trap which is a short gas chromatographic column that retards the flow of the compounds of interest at ambient temperature while venting the purge gas and, depending on the adsorbent used, much of the water vapor. The trap is constructed with a low thermal mass to allow rapid heating for efficient desorption, and rapid cooling to ambient temperature for recycling. The trap length, diameter, and wall thickness indicated in Figure 3.2 are critical and variations in these will affect the trapping and desorption efficiencies of the compounds discussed in this section.

The trapping and desorption efficiencies are also a function of the adsorbents, adsorbent mass, and the adsorbent packing order shown in

Figure 3.2. The single adsorbent Tenax GC (60/80 mesh) is effective for compounds that boil above approximately 30C. However compounds that boil below approximately 30C are not strongly adsorbed by Tenax and may be vented under the purging conditions. If compounds that boil below about 30C are to be measured, a dual adsorbent trap should be used. Grade-15 silica gel effectively retards the flow of most organics at ambient temperature and should be packed behind the Tenax to trap the lower boiling components. Silica gel is not a useful single adsorbent because higher boiling compounds do not efficiently desorb from it at 180C.

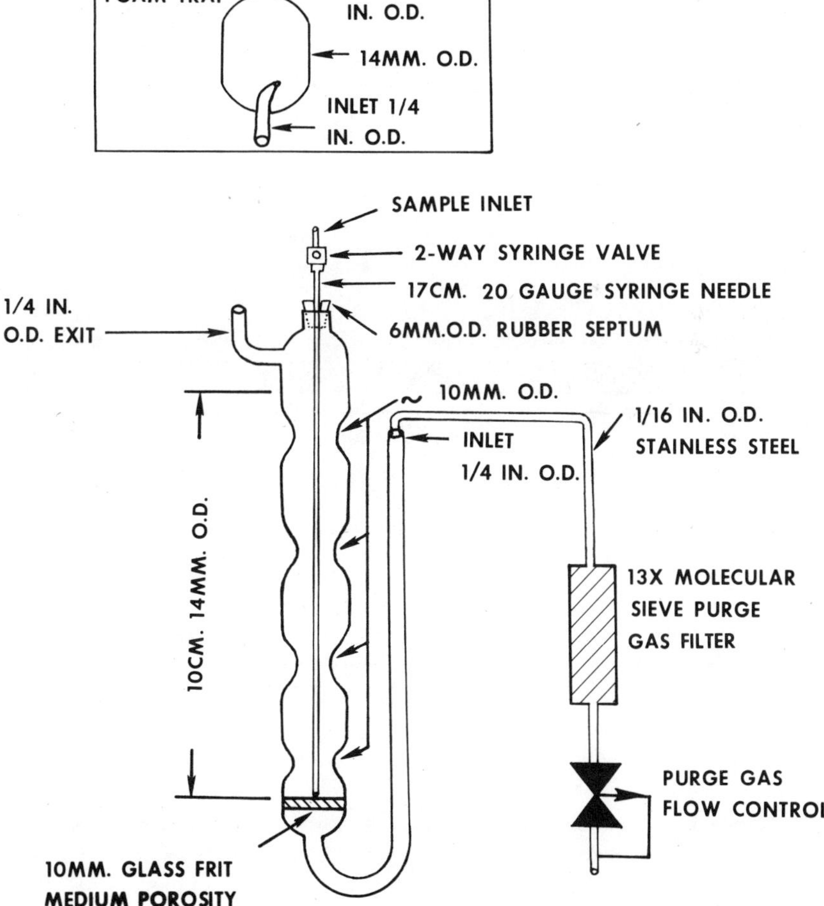

Figure 3.1 A Purging Device with a 5 ml Sample Capacity.

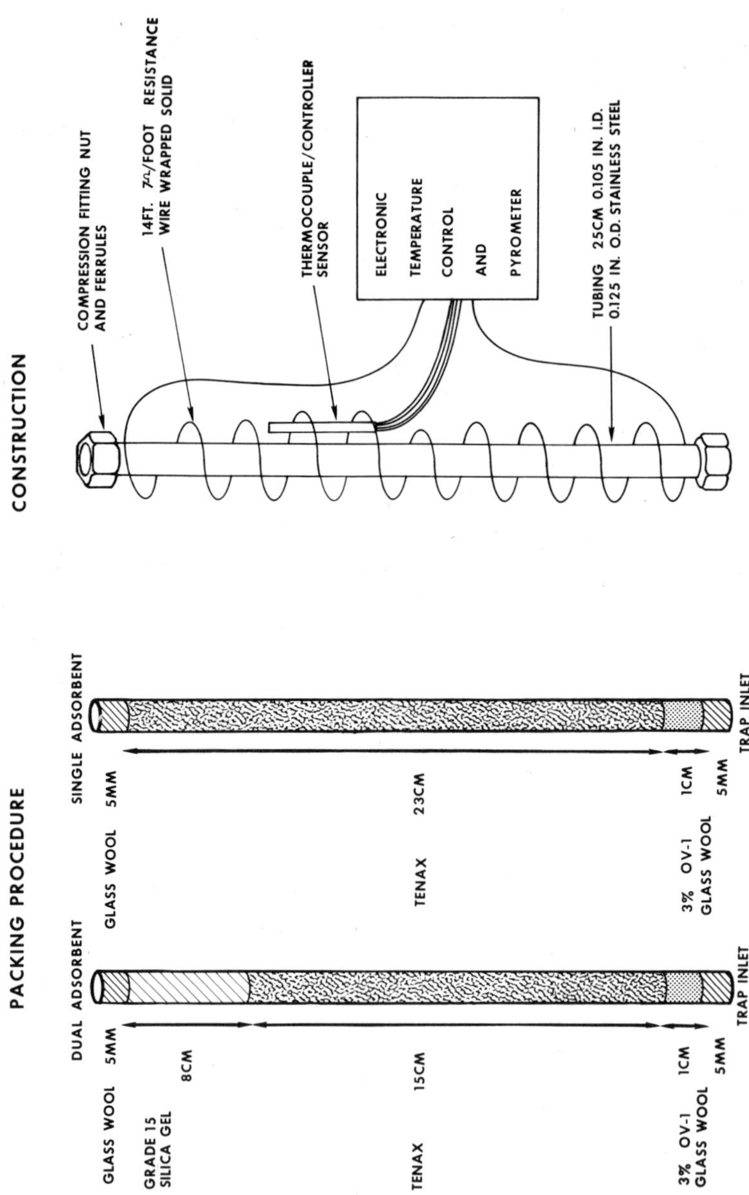

Figure 3.2 The Trap Assembly for a Purging Device

The Tenax-silica gel combination trap utilizes the adsorptive properties of two materials to provide a trap that effectively adsorbs and desorbs a wide variety of organic compounds. The small amount of OV-1 on glass wool at the trap inlet (Figure 3.2) is to insure that all the Tenax adsorbent is within the heated zone and is efficiently heated to the desorption temperature. A metal fitting at the trap inlet could act as a heat sink and create a cool spot on the Tenax if this spacer is not used.

Details of the trap heater are also shown in Figure 3.2. The adsorption-desorption cycle may be accomplished conveniently with the use of a six port valve and plumbing system constructed of materials that neither adsorb volatile organics nor outgas them. Several commercially available purge and trap systems use this approach. With the six port valve in the adsorb position, the effluent from the purging device passes through the trap where the flow of the organics is retarded and the purge gas is vented. During this period the gas chromatograph is supplied with carrier gas and may be used for other analyses. With the valve in the desorb position, the trap is placed in series with the gas chromatographic column which allows the carrier gas to back flush the trapped materials onto the chromatographic column.

It is strongly recommended that the power for the desorber heater be supplied by an electronic temperature controller that is set to begin supplying power as the valve is placed in the desorb position. This allows rapid heating of the trap to 180C with minimal overshoot and maintenance of the desorb temperature until desorption is complete (a four minute backflush at 20–60 ml/min is recommended). Using this procedure, the trapped compounds are released as a narrow plug into the gas chromatograph, which should be at the initial operating temperature. Packed columns with theoretical efficiencies near 500 plates/foot under programmed temperature conditions can usually accept such desorb injections without altering peak geometry.

Substitution of a non-controlled power supply, such as a manually operated variable transformer, will cause non-reproducible retention times and may lead to unreliable concentration measurements. If it is not possible to heat the trap in a rapid and controlled manner to the desorption temperature, the contents of the trap may be transferred onto the analytical column at 30C or lower and once again trapped. The analytical column is then rapidly heated to the initial operating temperature for the analysis.

Several gas chromatographic columns have been employed for the separation of the volatile components prior to measurement. A recommended column for general purpose work is a 8-ft x 0.1-in. id stainless steel or glass tube packed with 0.2% Carbowax 1500 on Carbopack-C (80/100 mesh). With a helium flow of 40 ml/min., the initial temperature of

60C is held for three min., then programmed at 8C/min. to 160C.
Newly packed traps should be conditioned overnight at 230C with an inert gas flow of at least 20 ml/min. The trap is also conditioned prior to daily use by backflushing at 180C for 10 min.

Discussion. Table 3.2 contains a list of some of the compounds that have been submitted to this type of analysis. The recovery data is intended to be illustrative only since recoveries depend strongly on several important method variables. Recoveries are expressed as a percentage of the amount added to organic free water. The purge time was 11–15 minutes with helium or nitrogen, the purge rate was 20 ml/minute at ambient temperature, and the trap was Tenax followed by Silica Gel. Data from the 5 ml sample was obtained with a custom made purging device and either flame ionization, microcoulometric, or electrolytic conductivity GC detectors. Data from the 25 ml sample was obtained with a Tekmar commercial liquid sample concentrator and a mass spectrometer GC detector using continuous repetitive measurement of spectra.

A number of other compounds have been concentrated and measured using the purge and trap method, but no recovery data for these is available. These compounds include chloromethane, bromomethane, chloroethane, 1,2-dichloropropane, trans-1,3-dichloropropene-1, cis-1,3-dichloropropene-1, 1,1,2-trichloroethane, 2-chloroethylvinylether, 1,1,2,2-tetrachloroethane, and ethylbenzene. In general the method is applicable to compounds that have a low solubility in water and a vapor pressure greater than water at ambient temperature.

All of the experiments summarized in Table 3.2 were conducted with the water sample at ambient temperature, about 22C. Purging at elevated temperatures has been investigated and clearly affects recoveries of some compounds. However, this is not recommended as a general procedure because the elevated temperature may promote chemical reactions that could significantly alter the composition of the trace organics. This is especially important with chlorinated drinking water or waste effluents.

A significant method variable is the purge gas flow rate. The total purge time is not very flexible since it will usually be desirable to keep this as short as possible to minimize the analysis time. In all of the experiments summarized in Table 3.2, a flow rate of 20 ml/min. was employed for 11–15 minutes, and for many compounds an acceptable recovery was obtained. Figure 3.3 shows the percentage recovery of several representative compounds as a function of purge gas flow rate. The general curve shape displayed for 1,2-dichloroethane and bromoform is typical of most of the compounds in Table 3.2. The low boiling compound vinyl chloride was trapped on Tenax only and its flow rate curve illustrates the sharp reduction in trapping efficiency observed with this type of compound and trap at elevated flow rates. The compound dichlorodifluoromethane displays a similar flow rate curve even with the combination Tenax-silica gel trap.

Because of the differences in the construction of various purge and trap devices, actual recoveries may vary significantly from those shown in Figure 3.3 and Table 3.2. Therefore, it is required that individual investigators determine recoveries of compounds to be measured as a function of flow rate with their apparatus. Operation in the optimum flow rate range will assure maximum sensitivity and precision for the compounds measured.

The recoveries of aliphatic hydrocarbons were found somewhat more variable than the recoveries of the other compounds investigated with this method. In all of these experiments, known quantities were added to organic free water, and the slightly soluble aliphatic hydrocarbons probably formed a thin surface layer on the water. Under these inhomogeneous conditions, special care is needed to achieve consistent results.

In order to generate quantitative measurements within a reasonable purge time, e.g. 10–15 min., calibration of the method with known

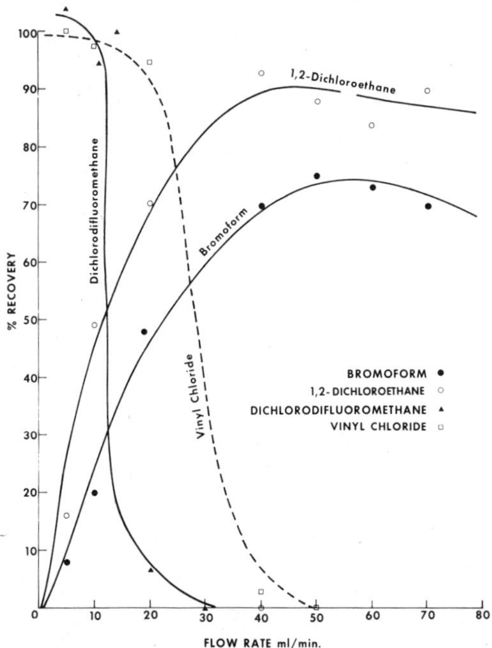

Figure 3.3 Recoveries of Selected Compounds as a Function of Purge Gas Flow Rate.

Table 3.2. Recoveries of Organics by Gas Purging and Trapping.

	Amount Added ug/l	% Recovery 5ml Sample	Amount Added ug/l	% Recovery 25 ml Sample
Chlorinated Hydrocarbons				
methylene chloride			12	97
chloroform	160	101	12	95
chlorodibromomethane	8,2	86,69	12	72
bromodichloromethane	40,2	94,82	12	72
carbontetrachloride	4,2	108,82		
1,2–dichloroethane	2	70	12	95
1,1,2–trichloroethylene	2,2	118,89		
tetrachloroethylene	2,2	81,80		
chlorobenzene	4	80		
p–dichlorobenzene	4	71	12	58
dichlorofluoromethane	2	7		
trichlorofluoromethane	2	99		
vinyl chloride	2	95		
1,1–dichloroethylene	2	114		
1,1–dichloroethane	2	90		

trans–1,2–dichloroethylene	2	100		
1,1,1–trichloroethane	2	88		
dichloroiodomethane	2	72		
2,3–dichloropropene–1	2	85		
Hydrocarbons				
n–pentane	81	100		
n–nonane	93	99		
n–pentadecane	100	56		
benzene			12	105
toluene			12	91
Brominated Hydrocarbons				
bromoform	4,2	67,48	12	73
1,2–dibromoethane	4,2	80,47		
Nitrogen Compounds				
nitromethane			24	2
nitrotrichloromethane			24	22
N–nitrosodimethylamine			24	0
N–nitrosodiethylamine			12	0
N–nitrosodi–n–butylamine			12	0

standards is required. The recommended approach is to estimate the concentration of the unknown by a comparison of its peak size with the size of the corresponding peak in a quality control check standard made at some appropriate concentration level and measured at regular intervals during the work day. From this estimate a concentration calibration standard is prepared with the concentrations of the compounds to be measured within a factor of two or less of the probable concentration in the unknown. Standards are prepared by taking aliquots of solutions in methanol and injecting them into organic free water. This insures maximum dispersal of the organic compound in the aqueous system. Organic free water is prepared by passing distilled water through an activated carbon column.

The standard is purged and measured immediately after the unknown and under the identical conditions used with the unknown. A sample of organic free water should be purged between each set of samples and especially after all high level standards or samples. This will insure that the apparatus is purged of contaminants and prevent cross contamination.

Sample matrices significantly different than surface water have not been investigated extensively. Extremes of pH, high ionic strength, or the presence of miscible organic solvents will likely affect recoveries of some compounds. Therefore measurements of compounds in these matrices must include determinations of recoveries of spikes in the sample matrix and perhaps use of a calibration based on the method of standard additions.

The detection limit of the method is also dependent on a number of operational variables. For a sample volume of 5 ml a concentration factor of about 1000 over a direct aqueous injection is usually possible. This places the limit of detection in the 0.1 to 1 microgram per liter range for a GC/MS system operating in the selected ion monitoring mode. With a GC/MS system operating in the continuous repetitive measurement of spectra mode, a sample volume of 25 ml is recommended to achieve this detection limit. The method can be applied over a concentration range of approximately 0.1 to 1500 micrograms per liter. Figure 3.4 shows a chromatogram of a mixture of 29 compounds from a purge and trap analysis.

Sample Collection and Preservation. Previous reports emphasized the importance of sample handling, and indeed because of the very volatile nature of the compounds measured in this type of analysis, sample collection deserves special consideration. In general, narrow mouth glass vials with a total volume in excess of 50 ml are acceptable. The bottles need not be rinsed or cleaned with organic solvents, but simply cleaned with detergent and water, rinsed with distilled water, air dried, and dried in a 105C oven for one hour. The vials are carefully filled with sample to overflowing (zero head space) and a Teflon faced silicone rubber septum is

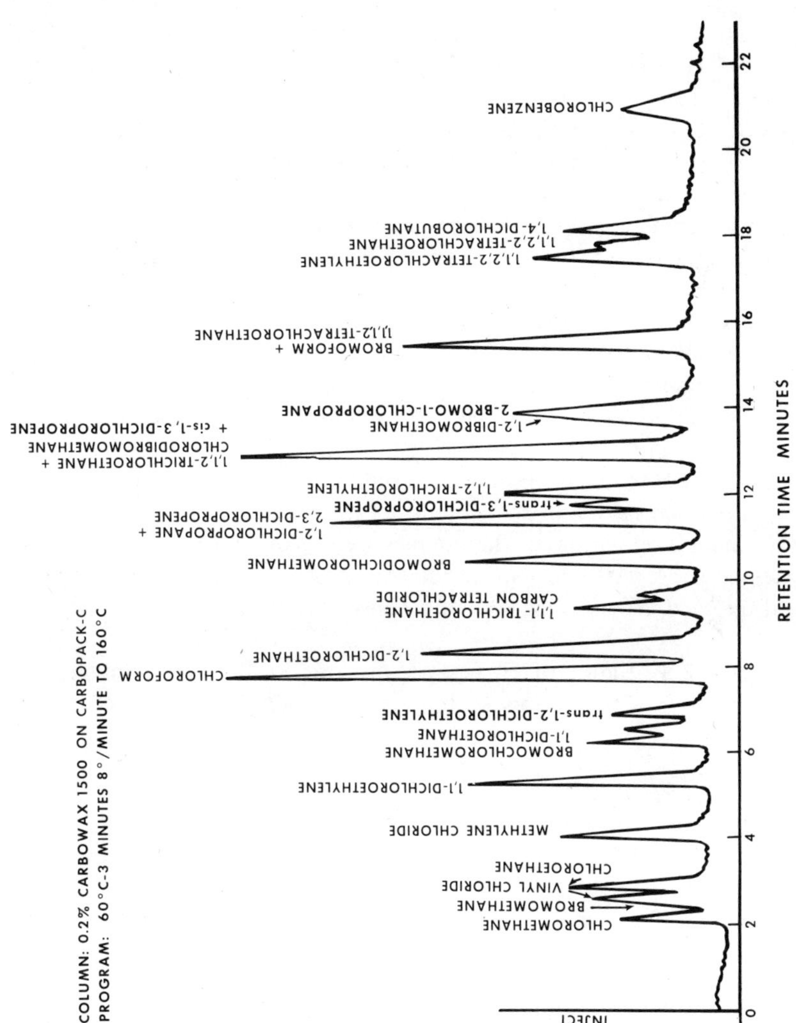

Figure 3.4 Chromatogram of Organohalides.

placed Teflon face down on the water sample surface. The septa may be cleaned in the same manner as the vials, but should not be heated more than one hour because the silicone layer slowly degrades at 105C.

Two types of seals for the vials have been employed and both give satisfactory results. Aluminum, one-piece, crimp-on seals used with serum vials and Teflon faced septa are acceptable if the seal is properly made and maintained during shipment. However, several years of experience indicates a success rate significantly less thant 100% in making proper seals of this type in the field. Therefore simple screw cap vials used with the Teflon faced septa were evaluated and found to give equivalent results and a very high rate of acceptable samples. Narrow mouth screw cap bottles with Teflon-faced silicone rubber septa cap liners are strongly recommended for sample collection.

One special problem in sample preservation has been recognized as a result of the widespread application of this method to drinking waters which contain residual quantities of disinfectants, e.g., chlorine. The levels of certain chlorinated compounds, e.g., chloroform, found in such waters will vary depending on the time of analysis unless the residual chlorine is consumed by a reducing agent such as sodium thiosulfate. Table 3.3 shows the concentrations of four compounds in Cincinnati tap water as a function of the sample age in days. The samples were taken from the distribution system at the EPA Environmental Research Center and maintained at 4C until analyzed. No reducing agent was added to the samples.

Table 3.3 Effect of Residual Chlorine on Concentrations of Chlorinated Methanes in Drinking Water at 4C

Time, days	$CHCl_3$ ug/l	$CHBrCl_2$ ug/l	$CHBr_2Cl$ ug/l	CCl_4 ug/l
0	17.0	12.4	11.9	3.1
1	17.9	12.9	11.7	3.1
2	23.7	16.1	14.4	3.4
7	32.6	19.9	17.2	3.4
8	36.3	21.6	18.7	3.5

The data in Table 3.3 shows that under the experimental conditions used the chloroform concentration increased by 114% during the eight day storage period. Smaller increases in the concentrations of the brominated compounds were observed, but the carbontetrachloride concentration did not change within the precision of the method which averages approximately 6% in the 1–1000 ug/l range. A similar set of samples was stored at 22C, and the same trends were observed except the concentrations of each of the halomethanes after seven and eight days were the same within the precision of the method. This indicates that the rate of halogenation was reduced substantially after one week of storage at 22C. The reduced rate of reaction was presumed due to a greatly diminished concentration of active halogenating agents, or the organic substrates, or both, but the details were not investigated further. In a third set of samples the reducing agent sodium thiosulfate was added at the time of sampling to quench the halogenation reaction. The concentration of the halomethanes in these remained constant within the precision of the method over the eight day period. In these samples no significant losses were observed indicating the effectiveness of the sample sealing procedures described in this paper.

The designer of an analytical survey that will include samples containing residual chlorine or other similar agents must be aware of these effects. The addition of a reducing agent at the time and place of sampling will give a measure of the instantaneous concentration of halogenated compounds, and this may or may not be the desired value. A measure of the maximum possible concentration of halogenated species in the particular sample may be obtained by storing the sample at 22C or higher until the concentration of halogenated compounds is constant.

Procedure. The following procedure will apply to both commercially available purging instruments and home-made units. However, home-made units usually demand trap manipulation which will not be discussed.

1. Adjust the helium flow through the purging vessel to 40 ml/min.
2. Newly packed traps must be conditioned at approximately 230C with a helium flow of 40 ml/min for 16 to 24 hours. Each day before use, the trap should be conditioned at 180C for 10 minutes during a backflush with helium at 30 ml/min.
3. Inject 25 ml of sample into a 25 ml purging device with a syringe. A sample volume of 25 ml is recommended when GC/MS is used for identification and measurement of the purgeable compounds. For more concentrated samples, a 5 ml volume can be used.
4. Purge the sample for 12 minutes at 40 ml per minute at room temperature.
5. After the 12 minute purge time, desorb the trapped compounds onto the GC column which has been cooled to room temperature for 10 minutes. Desorption is accomplished by rapidly heating the trap to 180C while backflushing with helium at 30 ml per minute for 4 minutes.

6. When the transfer is complete, heat the GC column rapidly to the initial operating temperature for analysis. Temperature programming at a suitable rate (4–8 degrees per minute) should be used to obtain sufficient resolution between GC peaks and accurate mass spectra.
7. Either the control mode or the IFSS mode may be used for data acquisition.

The sample dialogue below contains some suggested operating parameters.

```
SELECT MODE:  CONTROL
CALIBRATE?:  N
TITLE:  UP TO 64 CHARACTERS
CALIBRATION FILE NAME:  CURRENT FILE
FILE NAME:  UP TO SIX CHARACTERS
MASS RANGE:  20–260
INTEGRATION TIME:  17
SAMPLES/AMU:  1
THRESHOLD:  press return
RT ON CRT?:  N
RT GC ATTEN:  1 - 8
FAST SCAN OPT?:  N
MS RANGE SETTING?:  H
MAX RUN TIME:  30
DELAY BETWEEN SCANS (SECS.)?:  press return

SELECT MODE:  IFSS
CALIBRATE?:  N
TITLE:  UP TO 64 CHARACTERS
CALIBRATION FILE NAME:  CURRENT FILE
FILE NAME:  UP TO SIX CHARACTERS
MASS RANGE:  20–260
SAMPLES/AMU:  1
MAX RPT COUNT:  32
BASE INTEGRATION TIME:  1
RPT COUNT BEFORE CHECKING LOWER THRESHOLD:  8
LOWER THRESHOLD:  4
UPPER THRESHOLD:  4
RT ON CRT?:  N
RT GC ATTEN:  1 - 8
FAST SCAN OPT?:  N
MS RANGE SETTING?:  H
```

MAX RUN TIME: 30
DELAY BETWEEN SCANS (SECS.)?: press return

QUALITATIVE HEADSPACE ANALYSIS

Headspace analysis is a method of measurement for compounds that have a relatively high vapor pressure over a water sample. The primary purpose of the method is qualitative analysis because accurate concentration measurements require extensive information about the equilibria of many compounds between the vapor and dissolved phases. The principal advantage of the method is that it permits the rapid identification of very volatile compounds, e.g., vinyl chloride, fluorocarbons, etc., that may be missed with some other methods.

Samples should be collected in bottles as described in the inert gas purge and trap procedure. However, in place of the standard screw-on caps, the open top caps should be used. This permits sampling of the headspace without removing the screw-on cap. Glassware for blanks and samples should be prepared as described in the purge and trap procedure. In the laboratory, prior to sample collection, prepare a blank by completely filling a sample bottle with low-organic water. Low-organic water is prepared by passing distilled water through an activated carbon column and purging the carbon-treated water with an inert gas. The filled bottles are sealed and shipped to the sampling site along with the empty sample bottles. While samples are being taken, the blank bottles are opened and about 2 ml of water is decanted. The blank bottles are resealed and returned to the laboratory. These blanks will compensate for residual organics in the atmosphere at the sampling site as well as contamination during shipment and storage. Samples are taken by completely filling the bottles, then decanting about 2 ml. The bottles are capped, shipped at 0–5C, and the samples analyzed within 48 hours.

For the analysis, place the sample bottles in a water bath at 25C and allow temperature equilibration to standardize the headspace measurement. During this thermal equilibration, occasionally agitate the contents of the bottle to promote equilibrium between the gas and liquid phases. The GC columns suggested for direct aqueous injection in Table 2.7 are also recommended for the qualitative headspace analysis.

Flush a valve-controlled gas tight syringe twice with helium. To sample the headspace, pull 2 ml of helium into the syringe, close the valve, pierce the sample bottle septum, and inject the helium into the headspace, of the sample bottle. Without removing the needle from the septum, pull the syringe barrel out to the 5 ml mark, close the valve, and withdraw the syringe from the septum. To inject into the chromatograph, compress the gases in the syringe against the closed valve to about 2 ml, insert the needle into the injection block, and open the syringe valve. After completing the injection of a compressed plug of gas, close the valve and withdraw the syringe. Initial column temperature should be as near ambient as possible. A holding time of several minutes is suggested followed by temperature programming at 4-8 C/min. A sample run time of about 30 min is usually sufficient. Suggested sample data acquisition parameters are as follows:

```
SELECT MODE:  CONT
CALIBRATE?:  N
TITLE:  UP TO 64 CHARACTERS
CALIBRATION FILE NAME:  CURRENT FILE
FILE NAME:  UP TO 6 CHARACTERS
MASS RANGE:  14–16;19–27;29–31;33–260
INTEGRATION TIME  17;17;17;17
SAMPLES/AMU:  1;1;1;1
THRESHOLD:  press return
RT ON CRT?:  N
RT GC ATTEN:  1 - 8
FAST SCAN OPT?:  N
MS RANGE SETTING?:  H
MAX RUN TIME:  30
DELAY BETWEEN SCANS (SECS.)?:  press return
```

```
SELECT MODE:  IFSS
CALIBRATE?:  N
TITLE:  UP TO 64 CHARACTERS
CALIBRATION FILE NAME:  CURRENT FILE
FILE NAME:  UP TO 6 CHARACTERS
MASS RANGE:  14–16;19–27;29–31;33–260
SAMPLES/AMU:  1;1;1;1
MAX RPT COUNT:  32
BASE INTEGRATION TIME:  1
RPT BEFORE CHECKING LOWER THRESHOLD:  8
LOWER THRESHOLD:  4
UPPER THRESHOLD:  1
RT ON CRT?:  N
RT GC ATTEN:  1 - 8
FAST SCAN OPT?:  N
MS RANGE SETTING?:  H
MAX RUN TIME:  30
DELAY BETWEEN SCANS (SECS.)?:  press return
```

EXTRACTION WITH A LOW BOILING SOLVENT

Liquid-liquid extraction with a low boiling solvent is a time honored technique in organic analysis. The principal advantages of this method are its applicability to a broad variety of compounds and the large concentration factors that may be realized. A disadvantage is that very volatile compounds, e.g., chloroform, vinyl chloride, etc., will not be observed as they are either lost during extract concentration or masked during solvent elution from the GC.

The scope, limitations, and detection limits of the method are a function of the solvent selected. Over the years a great variety of solvents have been used. The most popular low boiling solvents for liquid-liquid extraction are diethyl ether, hexane, benzene, ethyl acetate, carbontetrachloride, chloroform, and methylene chloride. Each of these solvents may be the optimum for a given situation, but methylene chloride is the only solvent recommended for general purpose, broad spectrum extractions in this section. Among all the solvents available it has the maximum number of favorable properties.

Diethyl ether is a poor choice because it invariably contains peroxides and these must be removed before its use. However, air oxidation of diethyl ether produces more peroxides and, even immediately after purification, sufficient concentrations are present to react with trace organics and solvent molecules to create an unacceptable reagent blank. On balance diethyl ether, a hazardous flammable substance, is simply not of sufficient value to justify its continued use.

Hexane, benzene, and analogous hydrocarbon solvents are insufficiently polar to extract efficiently a wide variety of polar organics that are amenable to gas chromatography. This makes them unsuitable for the broad spectrum approach. Ethyl acetate is used in the biomedical field for extractions of drugs from body fluids, but it has not been evaluated sufficiently for broad spectrum applications in water analysis. All of the above solvents have the additional minor disadvantage of being lighter than water which is an inconvenience during separation of immiscible phases.

Carbontetrachloride is a carcinogenic material and its human inhalation toxic dose is 20 ppm (central nervous sytem effects). In addition its relatively low polarity, relatively high boiling point, and relatively high molecular weight make it clearly unsuitable for GC/MS work.

Chloroform is reported to be carcinogenic and its human inhalation toxic dose (systemic effects) is 10 ppm. By contrast, methylene chloride is not carcinogenic and the corresponding toxic dose for methylene chloride is 500 ppm (central nervous system effects). Of particular concern to GC/MS users is the tendency of both of these chlorinated solvents to pass through the sample/carrier gas enrichment device. This causes the development of excessive pressures in the mass spectrometer which may bring about an automatic shut-down. One solution is the use of a solvent venting valve. However if this is not available, the replacement of the chlorinated solvent during the concentration of the solvent extract is recommended. Methylene chloride (bp 41C) is readily replaced by acetone (bp 56C) but chloroform (bp 62C) is not. Up to 8 ul of acetone may be injected into the GC/MS without recourse to a venting valve. The lower boiling point of methylene chloride gives it a major advantage over chloroform in reducing losses of volatile organics during extract concentration. Finally chloroform is not suitable for methylations with diazomethane whereas methylene chloride is a preferred solvent (see section 3.5).

On balance, methylene chloride is the solvent of choice for low boiling solvent extractions. It is commercially available in very pure form, contains no peroxides, has a very low boiling point, a low toxicity, and a polarity sufficient to extract a wide variety of polar organics. It is very insoluble in water, easily replaced by acetone, and is generally recognized by practicing chemists as the equivalent of chloroform for solvent extractions. Some representative single laboratory recovery data of spiked organics in distilled water by solvent extraction with methylene chloride are shown in Table 3.4. Additional classes of compounds not included in Table 3.4 have been observed by liquid-liquid extraction with methylene chloride. These include aliphatic hydrocarbons, benzene derivatives, organic phosphorous pesticides, some carbamates, alcohols, aldehydes, ketones, some carboxylic acids, esters, nitriles, sulfur compounds, and many others.

The detection limits for methylene chloride extractions depend on the size of the water sample, the extent of concentration of the extract, the extraction efficiency of specific substances, and the GC/MS behavior of the specific substances. With a three liter water sample and concentration of the extract to 100 ul, the usual detection limit is in the range of 10-50 ng/1 (10-50 ppt) for compounds with a favorable extraction efficiency ($>80\%$) and favorable GC/MS behavior.

For accurate quantitative analysis as well as to obtain additional qualitative information, known quantities of compounds should be dissolved in acetone and aliquots mixed with sample water. This spiked sample is extracted according to the procedure to determine the amount recovered and to verify that the compound was extracted in the correct pH fraction.

For each set of samples a reagent blank is required. A reagent blank is defined as an experiment that uses all procedures, quantities of materials, and glassware used in sample preparation except that no water or aqueous solution is used. It is required even when contamination from glassware and reagents is well controlled. The reagent blank result is the documentation

Table 3.4. Recoveries of Organics by Solvent Extraction with Methylene Chloride

Compound Added	Spike Con., ug/l	pH[1]	% Recovery
Chlorinated Hydrocarbons			
Lindane	2	N	100
Heptachlor	2	N	90
DDD–p,p′	2	N	97
DDT–p,p′	2	N	98
Endrin	2	N	105
Toxaphene	2	N	106
Aroclor 1016	2	N	75
Aroclor 1254	2	N	80
hexachloroethane	50	N	45
p–dichlorobenzene	2	N	103
bis(2–chloroisopropyl)ether	25	N	84
bis(2–chloroethyl)ether	25	N	88
Chlordane (technical)	2	N	88
Aldrin	2	N	94
Heptachlor epoxide	2	N	95
DDE–p,p′	2	N	104
Dieldrin	2	N	84
Aroclor 1242	2	N	84
hexachlorobenzene	2	N	84
m–dichlorobenzene	2	N	70
hexachlorocyclopentadiene	2	N	55
hexachloro–1,3–butadiene	25	N	69
1–chloro–4–nitrobenzene	2	N	80
Phenols			
phenol	5,25	A,A+N	52,59
dimethylphenol	2	N	75
2,4,6,–trichlorophenol	2	N	53
pentachlorophenol	5	A	65
2–nitrophenol	25	N	52
o–chlorophenol	25	N	30
Polycyclic Aromatic Hydrocarbons			
naphthalene	0.5	N	86
acenaphthylene	25	N	92
acenaphthene	0.5,25	N,N	101,92
pyrene	0.5	N	104
fluoranthene	2	N	60

Table 3.4. Recoveries of Organics by Solvent Extraction with Methylene Chloride (continued).

Compound Added	Spike Con., ug/l	pH[1]	% Recovery
chrysene	2	N	73
benzo[a]pyrene	2	N	76
indeno[1,2,3–cd]pyrene	25	N	59
phenanthrene	2	N	72
anthracene	2	N	88
dibenzo[a,h]anthracene	25	N	104
benzo[g,h,i]perylene	25	N	100
benzo[a]anthracene	25	N	100
fluorene	2	N	87
Nitrogen Compounds			
pyridine	100	B	92
nitrobenzene	100	B	98
o–nitrotoluene	25	N	88
aniline	100	B	81
o–methylaniline	100	B	97
di–n–butylamine	100	B	20
benzidine	20	N	57
N–nitrosodimethylamine	20	A,B,N	0,0,0
N–nitrosodiethylamine	20	N	49
N–nitrosodibutylamine	20	N	73
2–benzothiazole	25	N	95
hydrazobenzene	25	N	84
Oxygen Compounds			
n–hexanol	25	N	64
camphor	25	N	89
alpha terpineol	25	N	82
isobutyric acid	25	A	5
isovaleric acid	25	A	17
n–octanoic acid	30	N,A	86,15
methyl palmitate	25	N,A	72,31
n–tributyl phosphate	25	N	105
di(2–ethylhexyl)phthalate	2	N	95
dimethyl phthalate	2	N	75
diethyl phthalate	2	N	87
di–n–butyl phthalate	2	N	92
isophorone	2	N	100

[1]N=pH 6–8; A=pH 2–4; B=pH 10–12.

that proves that good control was exercised, and it defines precisely the level of background that was beyond control.

Reagents and Equipment. Redistill the methylene chloride and acetone in an all-glass laboratory still before use. Mesityl oxide (bp 129C) and diacetone alcohol (bp 168C) are common contaminants in acetone. The entire volume of the still pot should not be distilled to prevent contamination of the distillate with these higher boiling impurities. Also the pot residue should be discarded frequently.

Distillation of solvents is particularly important for analyses of relatively clean water, for example, drinking water or water from a clean lake. Distillation of methylene chloride and acetone (if used) should be carried out immediately before use since these compounds begin to develop low level impurities within hours even when stored in the dark under nitrogen. For waste effluents, the best available commercial solvents may be adequate for use without redistillation. This should be verified by GC/MS analysis of a reagent blank prior to the extraction of the samples.

Purify the sodium sulfate by extraction with 50 volume percent methylene chloride - acetone in a soxhlet extractor for 8 hours. Air dry the sodium sulfate and heat in a 120C oven for 4 hours.

Standard laboratory glassware is required for this procedure. Extractions of 3 liter samples require a 6 liter separatory funnel with a Teflon stopcock. Kuderna-Danish (KD) glassware is used for the concentration of extracts. A standard set of KD glassware consists of a 3 ball Snyder column, a 500 ml evaporating flask, and a 10 ml receiver ampul graduated in 0.1 ml increments. An all glass laboratory still and a Soxhlet extractor are required for purification of reagents.

Procedure. The following procedure is scaled to a 3 liter sample. This is most appropriate for relatively clean water, for example drinking or clean lake water. For effluents and similar water containing higher concentrations, the quantities of sample and reagents should be scaled accordingly. This procedure was designed to isolate the neutral, acidic, and basic compounds in three separate extracts, and the data in Table 3.4 was obtained in this way. An alternative used in the EPA sampling and analysis procedure for screening of industrial effluents for priority pollutants is to combine the basic and neutral compounds in a single fraction by initial adjustment of the pH to 11 or greater. This reduces the workload by one-third in that two rather than three extracts must be processed, but other complications that may result from this short-cut are not well established. One observation is that emulsions may be more likely in basic than in neutral or acidic media.

Glassware for blanks, spikes, and samples should be washed with detergent, rinsed with tap water, rinsed with distilled water, air dried and heated in a muffle furnace at a minimum temperature of 300C for one hour. One gallon jugs for samples should be cleaned using the same procedure. However, these jugs are usually made of a soft glass and should not be heated as rigorously as Pyrex glassware. Heating to 300C for fifteen minutes is recommended. The sample jugs must be supplied with Teflon cap liners.

1. Measure the pH of the gallon sample and transfer 3 liters to a 6 liter separatory funnel. If the pH is less than 6 or greater than 8, adjust the pH to 6-8 with concentrated HCl or NaOH. Extract the sample with 125 ml of methylene chloride with vigorous shaking for 3 minutes. Allow the layers to separate, then drain the methylene chloride layer into a suitable collection flask. Repeat twice with 75 ml portions of methylene chloride and combine the three extracts.
   If an emulsion develops, a number of techniques may be applied to resolve the phase boundary. Additions of reagents or heating the mixture is not recommended as these actions may cause more problems than they solve. One effective emulsion breaking technique is to drain off the emulsion, swirl several times in the drain flask, and pour the mixture gently through the water in the separatory funnel. If this produces a partial methylene chloride layer, separate the clear methylene chloride, and pour the remaining emulsion through the water until a satisfactory separation of layers is accomplished.
2. Pour the combined extracts through two inches of anhydrous sodium sulfate in a 19 mm I.D. glass column. Collect the dried extract in a 500-ml Kuderna Danish (KD) flask fitted with a 10-ml ampul. A reagent blank, treated exactly like the sample extracts, should be run along with a sample or set of samples.
3. After the combined extract has filtered through the sodium sulfate, rinse the sodium sulfate with 50 ml of acetone. This is done to rinse any residual sample from the sodium sulfate, and to introduce a nonchlorinated solvent into the sample for GC/MS injection. If the GC/MS system is equipped with a solvent venting valve and it is known that methylene chloride injections are well tolerated, the acetone addition may not be necessary.
4. Adjust the pH of the water sample to 2 using concentrated HCl and repeat steps 1, 2, and 3 this time using 75 ml of methylene chloride for each extraction. However, if a derivatization procedure is to be employed, for example the methylation of the acid fraction to

convert carboxylic acids to methyl esters, see section 3.5 first.

5. Adjust the pH of the water sample to 12 using saturated NaOH solution and repeat steps 1, 2, and 3 as in step 4. Three sample extracts are now contained in three KD flasks: the neutral compounds extracted from a solution of pH 6-8, the acid compounds extracted from a solution of pH=2, and the basic compounds extracted from a solution of pH=12. The reagent blank is in a fourth KD flask.
6. Fit a 3 ball Snyder column to each Kuderna-Danish (KD) flask and concentrate the extracts on a steam bath to approximately 5 ml. After concentration the methylene chloride (bp=41) will be completely removed and the sample will be contained in acetone (bp=56). The extract can be further concentrated in the graduated ampul in a warm water bath under a stream of clean, dry air or nitrogen. Concentration of the extract to 100 ul or less is entirely possible with very low losses. However, successful completion of this final concentration requires tender loving care! The temperature of the water bath must be no higher than 50C, the stream of gas must be truly gentle, and the inside walls of the ampuls must be repeatedly rinsed with small quantities of pure solvent.
7. Install a conditioned column in the gas chromatograph. The column packing can be any one of a number of solid supports coated with a suitable silicone oil. Table 2.6 contains a number of suggested GC columns. Section 6.3 contains information on open tubular columns. Temperature programming should be utilized with a relatively low initial temperature to enhance resolution of more volatile components, and a final temperature that does not exceed the maximum operating temperature of the column.

   Inject an appropriate volume of extract into the system with the ionizer off and either vent the solvent or allow the solvent to be pumped from the manifold. After about two minutes, or when the pressure reaches 10^{-5} torr, turn on the ionizer, start data acquisition, and begin temperature programming. Halt data acquisition after about 40 minutes or until peaks cease to appear. Either the control mode or the IFSS mode may be utilized. The sample dialogue below contains some suggested operating parameters:

```
SELECT MODE:  CONTROL
CALIBRATE?:  N
TITLE:  ANY TITLE UP TO 64 CHARACTERS
```

CALIBRATION FILE NAME: CURRENT CALIBRATION FILE NAME
FILE NAME: UP TO SIX CHARACTERS
MASS RANGE: 40–450
INTEGRATION TIME: 8
SAMPLES/AMU: 1
THRESHOLD: press return
RT ON CRT?: N
RT GC ATTN: 1–8
FAST SCAN OPT?: N
MS RANGE SETTING?: H
MAX RUN TIME: 45
DELAY BETWEEN SCANS (SECS.)?: press return

SELECT MODE: IFSS
CALIBRATE?: N
TITLE: ANY TITLE UP TO 64 CHARACTERS
CALIBRATION FILE NAME: CURRENT FILE
FILE NAME: UP TO SIX CHARACTERS
MASS RANGE: 40–450
SAMPLES/AMU: 1
MAX RPT COUNT: 32
BASE INTEGRATION TIME: 1
RPT COUNT BEFORE CHECKING LOWER THRESHOLD: 8
LOWER THRESHOLD: 4
UPPER THRESHOLD: 4
RT ON CRT?: N
RT GC ATTEN: 1–8
FAST SCAN OPT?: N
MS RANGE SETTING?: H
MAX RUN TIME: 45
DELAY BETWEEN SCANS (SECS.)?: press return

EXTRACTION WITH A HIGH BOILING SOLVENT

Extraction with a high boiling solvent is the application of a relatively high molecular weight late eluting solvent to the conventional liquid-liquid extraction technique. This method has the advantage of permitting the analysis of volatile compounds that are usually lost during the concentration of a low boiling solvent extract or masked by the solvent during gas chromatography. The extraction is accomplished with a relatively small volume of high boiling solvent and a relatively large volume of water. The extract is analyzed without further concentration, and no volatile compounds are lost. The late eluting character of the solvent permits data acquisition to begin immediately after injection, and no volatile compounds are masked by the solvent.

The scope of the high boiling solvent extraction procedure is similar to inert gas purging and trapping. However, high boiling solvent extraction has been much less thoroughly evaluated. Very limited information is available about the types of compounds extracted with various high boiling solvents, detection limits, or the precision and accuracy of concentration measurements.

High boiling solvent extraction has the advantage that no special equipment is required. Under these circumstances the method should be of interest primarily for rapid qualitative analysis.

A number of solvents appear suitable for this application but hexadecane has been used most frequently. Hexadecane is available commercially in rather pure form, it is very insoluble in water, and it elutes very late on many GC columns. Hexadecane does have the disadvantage of low polarity and is best suited for the extraction of non-polar compounds and mixtures such as aliphatic hydrocarbons, benzene, toluene, the xylenes, gasoline, aviation jet fuel, kerosene, carbontetrachloride, chloroform, and related materials. The extraction efficiency will vary with different classes of compounds and any concentration measurements must be considered as minimum until recoveries of spikes clearly establish the capabilities of the method. The minimum detectable level for some compounds is about 1 ug/1 when analyzing a 10 ml extract of a one liter water sample. The extraction efficiency has been found to be 85-95% for halogenated methanes and aromatic hydrocarbons such as benzene, toluene, and xylenes.

As always, reagent blanks must be employed to establish that contamination is well controlled and the precise level of contamination that is beyond control. It should be verified by GC/MS that each batch of hexadecane or another solvent is free of significant quantities of interferring substances. Purify sodium chloride and sodium sulfate by extraction with 50 volume per cent methylene chloride - acetone in a soxhlet extractor for 8 hours. Air dry the solids and heat in a 120C oven for 4 hours.

Samples are collected in one-liter glass jars with Teflon lined caps. Extracts are collected in 15 ml vials with Teflon faced septa and aluminum crimp-on or screw cap seals. This is required to prevent loss of volatile compounds. Vials are available from the Pierce Chemical Company, Box 117, Rockford, Illinois 61105. No other special equipment is required

Glassware for blanks, spikes, and samples should be washed with detergent, rinsed with tap water, rinsed with distilled water, thoroughly dried, and heated in a muffle furnace for at least one hour at a minimum temperature of 200C.

Add 10 ml of hexadecane to a one liter sample bottle and collect about 950 ml of water sample. Cap the bottle and shake vigorously. If the water temperature is 18C or below, the hexadecane will solidify. In this case, wait until the hexadecane melts before shaking. Ship the sample to the laboratory without refrigeration.

If phenols or other low molecular weight acids are of interest, adjust the pH of the sample to 2 with concentrated sulfuric acid. Shake the sample well and transfer the contents to a 2 liter separatory funnel. Shake vigorously for 2 minutes and allow the layers to separate. If an emulsion occurs, add 5 grams of sodium chloride, reshake, and allow the layers to separate.

Drain off the water layer and measure its volume in a 1000 ml graduated cylinder. The added NaCl will not increase the volume by a significant amount. Drain the hexadecane through a small filter funnel packed with 2 grams of anhydrous sodium sulfate over glass wool into a 15 ml vial. Seal the vial by crimping on a Teflon faced septum and an aluminum seal or use a screw-on cap. The extract is now ready for GC/MS analysis.

Suggested gas chromatographic columns for high-boiling solvent extract analysis are listed in Table 2.6. Temperature programming from ambient to about 200C at 8 degrees per minute may be utilized. Begin with GC oven door open to achieve an initial temperature of about 50C. Turn on the ionizer, start data acquisition, and then inject 2-8 microliters of extract. Close the oven door and begin temperature programming. Halt data acquisition before elution of the hexadecane peak (generally about 10 minutes). Either the control mode or the IFSS mode may be used for data acquisition. The sample dialogue below contains some suggested operating parameters:

```
SELECT MODE: CONTROL
CALIBRATE?: N
TITLE: UP TO 64 CHARACTERS
CALIBRATION FILE NAME: CURRENT FILE
FILE NAME: UP TO SIX CHARACTERS
MASS RANGE: 20–260
INTEGRATION TIME: 8
SAMPLES/AMU: 1
THRESHOLD: press return
RT ON CRT?: N
RT GC ATTEN: 1–8
FAST SCAN OPT?: N
MS RANGE SETTING?: H
MAX RUN TIME: 20
DELAY BETWEEN SCANS (SECS.)?: press return
```

```
SELECT MODE:  IFSS
CALIBRATE?:  N
TITLE:  UP TO 64 CHARACTERS
CALIBRATION FILE NAME:  CURRENT FILE
FILE NAME:  UP TO SIX CHARACTERS
MASS RANGE:  20-260
SAMPLES/AMU:  1
MAX RPT COUNT:  32
BASE INTEGRATION TIME:  1
RPT COUNT BEFORE CHECKING LOWER THRESHOLD:  8
LOWER THRESHOLD:  4
UPPER THRESHOLD:  4
RT ON CRT?:  N
RT GC ATTEN:  1-8
FAST SCAN OPT?:  N
MS RANGE SETTING?:  H
MAX RUN TIME:  20
DELAY BETWEEN SCANS (SECS.)?:  press return
```

ADSORPTION WITH POROUS POLYMERS

Adsorption with porous polymers is the application of a solid phase, organic polymeric material to the isolation and concentration of organic compounds in water. The principal advantages of this technique are the acquisition of time integrated (composite) samples and the very low detection limits made possible by sampling relatively large volumes of water.

In the past, activated carbon has been used extensively as an adsorbent for organic compounds in water. However the carbon adsorption method, although still widely used, has a number of limitations. High quality activated carbon is not always readily available, significant effort is required to reduce background, lengthy work-up procedures are required, and many organic compounds are adsorbed so strongly that desorption is accomplished in very low yield.

The porous polymer adsorption method has the same advantages as the carbon adsorption method, but fewer disadvantages. A variety of commercial polymers are available, background contamination is more easily removed, desorption does not require lengthy Soxhlet extraction, and the recoveries in Table 3.5 are acceptable for measurements of a wide variety of compounds. These recoveries are based on elution with diethyl ether and were selected from the literature cited in section 10.3 by Robert Kleopfer of EPA.

Table 3.5. Recoveries of Organics by Adsorption on Porous Polymers.

Compound Added	XAD Resin Type	%Recovery
Chlorinated Hydrocarbons		
sym–tetrachloroethane	2, 4, 7, 8	61, 72, 35, 59
bis–chloroisopropyl ether	2, 4, 7, 8	76, 80, 76, 77
bromodichloromethane	2	87
dibromochloromethane	2	99
chlorobenzene	2	95
o–dichlorobenzene	2	88
m–dichlorobenzene	2	93
1,2,4,5–tetrachlorobenzene	2	74
1,2,4–trichlorobenzene	2	99
benzyl chloride	2	88
o–chlorobenzyl chloride	2	96
m–chlorotoluene	2	80
2,4–dichlorotoluene	2	71
p–chloronitrobenzene	2	100
Lindane	2	95
Aldrin	2	47
Dieldrin	2	93
Chlordane	2	82
DDT	2	96
DDE	2	81
Phenols		
phenol	2, 4, 7, 8	14, 40, 19, 29
o–cresol	4	73
p–cresol	2, 4, 7, 8	44, 69, 33, 47
3,5–xylenol	4	79
o–chlorophenol	4	96
p–chlorophenol	4	95
2,4,6–trichlorophenol	4	99
pentachlorophenol	2, 4, 7, 8	84, 84, 83, 77
1–naphthol	4	91
Polycyclic Aromatic Hydrocarbons		
naphthalene	2, 4, 7, 8	79, 80, 64, 78
1–methylnaphthalene	2, 4, 7, 8	87, 77, 64, 80
2–methylnaphthalene	2, 4, 7, 8	95, 77, 63, 77
2–methoxynaphthalene	2	97
acenaphthene	2, 4, 7, 8	99, 81, 72, 20

Table 3.5. Recoveries of Organics by Adsorption on Porous Polymers. (continued)

Compound Added	XAD Resin Type	%Recovery
biphenyl	2	101
fluorene	2	84
anthracene	2	83
tetrahydronaphthalene	2	62
dibenzofuran	2, 4, 7, 8	93, 82, 73, 95
diphenyl ether	2	91
Nitrogen Compounds		
o–nitrotoluene	2, 4, 7, 8	82, 83, 53, 77
nitrobenzene	2	91
o–nitroaniline	2	100
hexadecylamine	2	94
indole	2	89
N–methylaniline	2	84
quinoline	2	84
isoquinoline	2	83
benzonitrile	2	88
Atrazine	2, 2	83, 100
benzothiazole	2, 4, 7, 8	100, 82, 40, 53
benzoxazole	2	92
Carboxylic Acids		
octanoic acid	4	108
decanoic acid	4	90
palmitic acid	4	101
oleic acid	4	100
benzoic acid	4	107
Alcohols		
1–hexanol	2	93
2–octanol	2	100
1–decanol	2	91
benzyl alcohol	2	91
cinnamyl alcohol	2	85
2–phenoxyethanol	2	102
alpha terpineol	2, 4, 7, 8	81, 80, 36, 62
2–ethylhexanol	2, 4, 7, 8	99, 91, 74, 79
1–dodecanol	2	93

Table 3.5. Recoveries of Organics by Adsorption on Porous Polymers. (continued)

Compound Added	XAD Resin Type	%Recovery
Aldehydes and Ketones		
2,6–dimethyl–4–heptanone	2	93
2–undecanone	2	88
acetophenone	2	92
benzophenone	2	93
benzil	2	97
benzaldehyde	2	101
salicylaldehyde	2	100
isophorone	2, 4, 7, 8	76, 86, 46, 47
Esters		
benzyl acetate	2	100
di(methoxyethyl)phthalate	2	94
dimethyl phthalate	2	91
diethyl phthalate	2	92
dibutyl phthalate	2	99
di(2–ethylhexyl)phthalate	2	88
diethyl fumarate	2	86
dibutyl fumarate	2	92
di(2–ethylhexyl)fumarate	2	84
diethyl malonate	2	103
methyl benzoate	2	101
methyl decanoate	2	95
methyl octanoate	2	98
methyl palmitate	2	70
methyl salicylate	2	96
methyl methacrylate	2	35
Miscellaneous		
n–hexadecane	2, 4, 7, 8	3, 11, 3, 18
ethylbenzene	2	81
cumene	2	93
p–cymene	2	92
dihexylether	2	75
dibenzylether	2	99
anisole	2	87
bromoform	2	101
iodobenzene	2	81

The porous polymers employed in this method are called macroreticular resins. Macroreticular refers to the relatively large, controlled pore size of the resin beads. Each grain of resin is formed from many microbeads that are cemented together during the polymerization process. The pores are the spaces left between the cemented microspheres.

The most-frequently used resins are the Amberlite XAD series made by the Rohm and Haas Company. They are hard insoluble beads of 20-50 mesh, varying from white to light brown in color. Four materials are available, XAD-2, 4, 7 and 8. XAD-2 and XAD-4 are styrene-divinylbenzene copolymers that are chemically identical but differ in active surface area and average pore diameter. XAD-2 has 300 m^2/g active surface area and 90 A average pore diameter compared to 784 m^2/g and 50A for XAD-4. XAD-7 and XAD-8 are both acrylate polymers having active surface areas of 750 m^2/g and 140 m^2/g and average pore diameters of 80 A and 250 A, respectively. The resins derive their adsorptive properties from their macroreticular porosity, pore size distribution, and high surface area. XAD's 2, 4, and 7 are available from Mallinckrodt distributors such as Curtin. All four are available from Chemical Dynamics Corp., Hadley Road, P.O. Box 395, South Plainfield, New Jersey 07080.

In practice, water is passed through a column of XAD resin and the organic materials are adsorbed. The organics are desorbed from the resin by elution with an organic solvent and the extract is submitted to GC/MS analysis. A number of elution solvents have been evaluated and methylene chloride and acetone are very effective. Diethyl ether and chloroform are not acceptable for the reasons discussed in the low boiling solvent extraction method.

The method has been applied to a wide variety of organic compounds and is applicable to non-ionized materials having boiling points above about 100C. The detection limit is of the order of 1-10 ng/1 in an aqueous matrix. In general, overall recoveries are equivalent to those obtained with low boiling solvent extraction. Table 3.5 contains a number of values that have been reported using diethyl ether for elution. For accurate concentration measurements, these recoveries must be established in the laboratory conducting the procedure.

As in other methods, a reagent blank is required to assure that compounds detected are not contaminants of the glassware or materials. The reagent blank may consist of one liter of low organic water.

Reagents and Equipment. Extract glass wool in a Soxhlet apparatus with methylene chloride before use. Methanol, methylene chloride, and acetone should be of high quality and redistilled before use in all-glass apparatus. (See the low boiling solvent extraction method for precautions in the use of these solvents). Sodium sulfate should be extracted and dried before use.

Low organic water is prepared by passing tap water through a filtration system consisting of cation and anion-exchange columns and activated carbon columns. This water is then distilled in an all-glass apparatus. The resulting deionized/distilled water is then passed through an XAD-2 column as a final clean-up step.

Some special equipment is required for this method. The resin columns are custom made from liquid chromatography columns, 25 mm O.D., 22 mm I.D., equipped with a Teflon stopcock. The columns are shortened to about 180 mm (packing length) and fitted with 24/40 female joints at the top of the column. The column also should have glass "ears" near the top for securing the glass stoppers or the glass adapters with rubber bands or wire springs. Other arrangements are also acceptable; the main requirement is that the ratio of length to inside diameter should be about 7. The stoppers are 24/40 ground glass stoppers equipped with glass "ears".

For automatic continuous sampling, a variable speed peristaltic pump capable of pumping at a rate of 120 ml/min is used. The pump is attached to the stopcock end of the column with Tygon tubing. The water sample is drawn through the column from the other end using an adapter and Teflon tubing. The recommended flow rate is four bed volumes per minute.

Adapters are required to connect the column with 1/4 inch O.D. Teflon tubing that is placed in the sample source. The adapter consists of a 24/40 male joint joined to a short section of 1/4 O.D. glass tubing. The glass tubing is connected to the Teflon tubing by a brass Swagelok adapter fitted with Teflon ferrules. The glass adapter is equipped with glass "ears" for securing in place rubber bands or wire springs.

Kuderna-Danish equipment is required for the final concentration of the eluate. (See the low boiling solvent extraction procedure for a list of this equipment).

Procedure. The commercial XAD resins are contaminated. Preextraction to remove interferring impurities is required.

1. Slurry the desired amount of resin in methanol and decant to remove the fine particles. Place the resin in a Soxhlet apparatus of appropriate size using glass wool plugs to hold the resin in place.
2. Extract the resin sequentially with acetone, methanol, and methylene chloride utilizing a minimum of ten cycles per solvent.
3. With the resin still in the Soxhlet apparatus, rinse three times with methanol. Store the resin in a closed glass container under methanol. Do not allow to dry.

   An alternative cleanup procedure uses extraction with acetone and acetonitrile, followed by vacuum degassing at 200C and 10^{-6} to 10^{-7} torr. If a large Soxhlet is not available, 100 g batches can be cleaned up by placing them in a 5 cm I.D. column and washing successively with two liters of acetone, methanol, and methylene chloride. Concentrate a portion of the methylene chloride effluent, and check it for background. Repeat the methylene chloride wash if necessary. Remove the resin from the column and store under methanol.

4. Plug the bottom of the resin column with 1-2 cm of pre-extracted glass wool. Add the XAD resin to the column as a methanol slurry. Adjust the resin height in the column to about 8 cm. Stir with a glass rod to remove any trapped air. Plug the top of the column with 1-2 cm of glass wool. Do not allow the column to dry.
5. Add 30 ml of low organic water, let drain until water reaches the top of the glass wool. Repeat three times.
6. The column is now ready for a sample. If stored or shipped, the column should be filled with water and stoppered such that all air pockets are excluded.
7. Grab samples generally consist of 1 to 5 liters of water. It may be desirable to adjust the pH of the sample prior to passing it through the column. For example, the recovery efficiencies for phenols and carboxylic acids are greater when performed at lower pH. Pass the water through columns by gravity at a flow rate of approximately four bed volumes per minute (ca. 120 ml/min.). Any sample except drinking water may contain suspended matter that is trapped on the glass wool and reduces the flow. In this case a peristaltic pump at the bottom of the column for samples greater than 1 liter is suggested. Allow sample to drain freely and close the stopcock. If the column is to be stored or shipped prior to solvent extraction, fill with organic free water, exclude air bubbles, and stopper. Frequently air bubbles form in the column during the adsorption step; ignore them.
8. For composite samples, the columns can be used for manual compositing. For example, in the field, one liter of effluent water can be passed through the column every hour over the desired sampling period (8-24 hours) as a time composite. Composites can also be taken by attaching the column to a peristaltic pump having the appropriate pumping speed (ca. 120 ml/min or less). The water to be sampled is drawn to the column through Teflon tubing (1/4 inch O.D.). The effluent end of the column is attached to the peristaltic pump with Tygon tubing. The flow rate should be checked periodically to determine the volume sampled.
9. For desorption, allow the column to drain freely. Close the stopcock and place a 300 ml Erlenmeyer flask under the column.

10. Add 30 ml of acetone and force the solvent through the resin at a slow rate (dropwise) using nitrogen pressure (ca. 2 p.s.i.) if necessary.
11. Allow the acetone to drain until it is just to the top of the upper glass wool plug.
12. Repeat steps 10 and 11.
13. Continue desorption with eight bed volumes of methylene chloride.
14. Remove the water layer from the eluate using a disposable glass pipette.
15. Dry the eluate with organic-free anhydrous sodium sulfate, and concentrate by Kuderna-Danish evaporation to a volume of about 5 ml. The volume can be further reduced over a warm water bath by directing a stream of clean air or nitrogen onto the extract. (See the low boiling solvent extraction procedure for cautions during concentration). The extract is now ready for GC/MS analysis.
16. Often the resin columns can be reused after a minimum amount of cleanup. However, if the sample was extremely dirty and contained sediment, it is recommended that the resin be discarded. If the resin is discolored, replace the upper glass wool plug. Add 30 ml of acetone and allow to drain freely. Repeat the acetone wash twice. Add 30 ml of organic free water and allow to drain freely. Repeat the water wash twice. The column is now ready to reuse.

    The GC/MS conditions are the same as given under the low boiling solvent extraction procedure. Suggested columns are in Table 2.6.

3.2 AIR SAMPLES

Air sampling and analysis methods have been under development for many years, and numerous methods are available for the isolation, concentration, and measurement of specific organic compounds in air samples. However, as emphasized in chapter one, the purpose of this manual is to describe broad spectrum methods for general classes of organics. In this area there is relatively little information available. Therefore, the methods given in this section probably represent the first generation broad spectrum air procedures, and significant improvements will probably occur in future years.

The problem of quantitative sampling of air is significantly more complex than is quantitative sampling of water. Numerous studies of the sampling problem have been conducted and the results reported. It is beyond the scope of this manual to reproduce the detailed methodology required for accurate quantitative sampling of organics in air. The broad spectrum methods described in this chapter are most appropriate for qualitative or rough quantitative analyses of air samples.

After sampling and the initial sample preparation, the GC/MS aspects of air analyses are not unlike the analyses of samples from other media. Therefore, a great deal of information from the other sections of this chapter and other chapters is applicable to air analyses. Numerous references to these sections are made under Air Samples.

DIRECT AIR INJECTION

Direct air injection is the introducion of a few ml, e.g., 5, of air into the GC/MS system. This procedure is clearly similar to the qualitative headspace analysis described in section 3.1. The advantage of this method is the speed of the analysis and relative simplicity of sample handling. The disadvantage is that the method is limited to relatively high concentrations of volatile compounds such as vinyl chloride or chlorotrifluoromethane. The detection limit is estimated to be about two milligrams per cubic meter (kiloliter) of air. This procedure will probably be useful only for an industrial work site or a spill site. Generally ambient air levels of organics are in range of tens of picograms to nanograms per cubic meter. Industrial work rooms or source effluents usually do not exceed tens of micrograms per cubic meter, but many exceptions are possible and have been reported. The application of selected ion monitoring as described in section 6.1 may lower the detection limit of the method to the microgram per cubic meter level.

For this procedure glass collecting tubes of 100–500 ml capacity fitted with gas tight stopcocks and rubber septa are evacuated in the laboratory. The stopcock is opened at the sampling site, and sampling is continued until the pressure in the collecting tube is equivalent to the external pressure. The stopcock is closed and the container is returned to the laboratory. An aliquot of the sample is removed with a gas tight syringe and injected directly into the GC/MC system Sampling containers are commonly available from general laboratory supply companies.

The GC columns suggested for direct aqueous injection in Table 2.7 are also recommended for direct air injection. Initial column temperature should be as near ambient as possible. A holding time of several minutes is suggested followed by temperature programming at 4–8 C/min. A sample run time of about 30 min. is usually sufficient. Suggested sample data acquisition parameters are as follows:

```
SELECT MODE:  CONT
CALIBRATE?:  N
TITLE:  UP TO 64 CHARACTERS
CALIBRATION FILE NAME:  CURRENT FILE
FILE NAME:  UP TO 6 CHARACTERS
MASS RANGE:  14-16;19-27;29-31;33-260
INTEGRATION TIME:  17;17;17;17
SAMPLES/AMU:  1;1;1;1
THRESHOLD:  press return
RT ON CRT?:  N
RT GC ATTEN:  1 - 8
FAST SCAN OPT?:  N
MS RANGE SETTING?:  H
MAX RUN TIME:  30
DELAY BETWEEN SCANS (SECS.)?:  press return

SELECT MODE:  IFSS
CALIBRATE?:  N
TITLE:  UP TO 64 CHARACTERS
CALIBRATION FILE NAME:  CURRENT FILE
FILE NAME:  UP TO 6 CHARACTERS
MASS RANGE:  14-16;19-27;29-31;33-260
SAMPLES?AMU:  1;1;1;1
MAX RPT COUNT:  32
BASE INTEGRATION TIME:  1
RPT BEFORE CHECKING LOWER THRESHOLD:  8
LOWER THRESHOLD:  4
UPPER THRESHOLD:  1
RT ON CRT?:  N
RT GC ATTEN  1 - 8
FAST SCAN OPT?:  N
MS RANGE SETTING?:  H
MAX RUN TIME:  30
DELAY BETWEEN SCANS (SECS.)?:  press return
```

ADSORPTION WITH POROUS POLYMERS

This method involves passing a measured amount of air through a solid adsorbent, followed by thermal desorption of organics into the GC/MS system. The advantage of this method compared to cryogenic trapping is the significantly simpler field equipment handling. The advantage, compared to solvent desorption, is higher sensitivity because the total sample is measured and there is no background from the solvent. The principal disadvantage is that because the total sample is measured, there is no room for error in the analytical procedure. Multiple samples should be collected to provide the required back-up in the event of a laboratory accident or equipment malfunction, and to obtain information on the precision of the method. The porous polymer adsorption method is applicable to organics having compositions in the C6 to C18 range.

Although several different adsorbents have been used by various researchers, Tenax GC appears to have several advantages over the others. This material is thermally stable to about 350C, has good adsorption characteristics, but does not efficiently retain water vapor. However caution must be excercised because some batches of Tenax have been reported to decompose at lower temperatures. Detection limits are not well established for these methods, but probably are about 1–10 ug per cubic meter (kiloliter) of air. Good quality control requires that unused adsorbent be desorbed at frequent intervals with the identical conditions used with the samples. This will assure that background contamination from the adsorbent itself or from the equipment is under control, or, at least, is known to the investigator. There is limited recovery data available for this method.

Tenax GC (2,6-diphenyl-p-phenyleneoxide polymer) is available from several suppliers including Applied Science, State College, Pennsylvania. This must be purified before use by extracting with methanol for several or more hours in a Soxhlet apparatus. Purification may not be necessary if purified material is available from the supplier, but the material is susceptible to decay and contamination during storage.

The Tenax GC, 60/80 mesh adsorbent is packed into Pyrex glass tubes of 2.5 mm I.D. x 200 mm dimensions. The adsorbent is supported by 0.5 cm plugs of silanized glass wool. Prior to use the tubes and glass wool may be heated to 500 C for several hours to remove traces of organic matter. The packed sampling tubes are conditioned at 270 – 300 C for 0.5 – 1 hour under a helium or nitrogen flow of 30 – 50 ml/min. After cooling to room temperature, cap both ends or store the sampling tube in a clean screwcapped test tube.

Sampling pumps are available from several laboratory supply companies. A flowmeter should be precalibrated with a sample tube connected using a soap bubble flowmeter. The sample is taken by pulling air through the tube at a rate of about 500 ml/min. After a suitable sampling period of 2 – 4 hours in urban locations, the pump is turned off, and the tube is capped or stored until the analysis can be performed.

The sample is analyzed by direct heat desorption at 250 – 270 C for three minutes under a 10 ml helium flow. A desorption technique similar to that described under inert gas purging and trapping is recommended. The GC column oven is reduced to ambient temperature before desorption begins, then programmed to an elevated temperature at 4 – 8/min. The columns described in Table 2.6 for solvent extracts are recommended for porous polymer trapped samples. Data collection should begin just prior to the desorption step. The following GC/MS operating conditions are recommended for the control and IFSS modes.

```
SYSTEM 1500 IS ON SELECT MODE:  CONT
CALIBRATE?:  N
TITLE:  UP TO 64 CHARACTERS
CALIBRATION FILE NAME:  CURRENT FILE
FILE NAME:  UP TO 64 CHARACTERS
MASS RANGE:  14-16;19-27;29-31;33-260
INTEGRATION TIME:  17;17;17;17
SAMPLES/AMU:  1;1;1;1
THRESHOLD:  press return
RT ON CRT?:  N
RT GC ATTEN:  1 - 8
FAST SCAN OPT?:  N
MS RANGE SETTING?:  H
MAX RUN TIME:  30
DELAY BETWEEN SCANS (SECS.)?:  press return

SELECT MODE:  IFSS
CALIBRATE?:  N
TITLE:  UP TO 64 CHARACTERS
CALIBRATION FILE NAME:  CURRENT FILE
FILE NAME:  UP TO 6 CHARACTERS
MASS RANGE:  14-16;19-27;29-31;33-260
SAMPLES/AMU:  1;1;1;1
MAX RPT COUNT:  32   BASE INTEGRATION TIME:  1
RPT BEFORE CHECKING LOWER THRESHOLD:  8
LOWER THRESHOLD:  4
UPPER THRESHOLD:  1
RT ON CRT?:  N
RT GC ATTEN:  1 - 8
FAST SCAN OPT?:  N
MS RANGE SETTING?:  H
MAX RUN TIME:  30
DELAY BETWEEN SCANS (SECS.)?:  press return
```

FILTRATION WITH GLASS FIBER FILTERS

This method is for the analysis of organic compounds contained in particulate matter (dirt, dust, etc.) collected from air with a glass fiber filter. Particulate matter itself is a criteria pollutant that is known to be rich in organic material. The advantage of this method is that it specifically addresses the problem of organics in particulate matter. The disadvantage is that gas phase, low molecular weight organics are not adsorbed on glass fibers. In this method air is drawn into a covered housing and through a glass fiber filter by means of high flow rate blower (HI VOL sampler), the mass of the collected particulate matter is determined, the organics are removed by Soxhlet extraction, and analyzed by GC/MS. The extract may be separated with silica gel into aliphatic, aromatic, and oxygenated fractions prior to GC/MS analysis. Generally, for urban particulates, alkanes ranging from $C_{17}H_{36}$ to $C_{31}H_{64}$ will be found in the aliphatic fraction, compounds such as anthracene, pyrene, and benzopyrene will be found in the aromatic fraction, and phenols, phthalate esters and other polar compounds will be found in the oxygenated fraction.

Methylene chloride is the solvent of choice for general purpose work. The reasons for choosing methylene chloride are exactly the same as given under the low boiling solvent extraction method in section 3.1 of this chapter. The specifications for glass fiber filters and the detailed procedure for use of a HI VOL sampler are beyond the scope of this manual, but are well documented in other publications. The user should see the *Federal Register*,36, 8186 (1971). Additional information is available from the director of the Environmental Monitoring and Support Laboratory, Research Triangle Park, North Carolina.

The detection limit for this method is not well established, but is probably about one ng per kiloliter (cubic meter) of air. Typically a 24 hour sample of 2500 kl in an urban location will yield 125 – 150 mg of particulate matter (60 – 200 ug/kl) of which about 10% is extractable. It is absolutely essential that an unused glass filter from the same batch of filters be extracted at the same time and under the same conditions as the loaded filters. The analysis of this unused filter will provide information about background and solvent contamination. There are no recovery data available for this method.

The methylene chloride should be distilled as described under low boiling solvent extraction immediately before use. The weighed filter is cut into pieces small enough to fit into a 250 ml Soxhlet extractor. Usually only one section of the whole filter is used for the organics analysis. No thimble should be used, but about one inch of pre-extracted glass wool is placed in the bottom of the extraction chamber. This will act as a filter and preclude the accumulation of particulate matter in the distillation pot. The filter pieces are inserted, and 200 ml of solvent is placed in the pot. Extraction should continue for 25 cycles, and the cooled extract transferred to a Kuderna-Danish (K-D) apparatus. If the optional silica gel separation is to be used, concentrate the extract to 1 ml; if no silica gel separation is planned, concentrate to 0.1 – 1 ml.

For the silica gel separation, pack activated silica gel into a glass chromatography column (2 cm I.D.) to a height of 10 cm. To the 1 ml of extract in a small beaker add sufficient silica gel to adsorb the sample, and then evaporate the residual solvent with gentle stirring in a hood. Wet the column with about 20 ml of hexane and add the sample when the last of the 20 ml reaches the surface of the adsorbent. Rinse the beaker with hexane and add the rinsings to the column. The beaker should be rinsed several times with hexane and each succeeding eluent when the eluent is added to the column. Collect the fractions in K-D ampuls and elute the aliphatic fraction with 85 ml of hexane, the aromatic fraction with 85 ml of benzene or the safer toluene, and the polar fraction with 1:1 methanol-methylene chloride. Concentrate each fraction to 0.1 – 1 ml in the K-D apparatus, and proceed with the GC/MS analysis of each.

The GC columns and conditions recommended are the same as those used for low boiling solvent extraction. This procedure should be consulted for this information. Also see the previous section for sample GC/MS system dialogue.

3.3 SEDIMENT SAMPLES

Sediment is defined as wet solids taken from the bottom of a stream, river or lake. This method is applicable to these, and with the elimination of the drying step, to the broad spectrum analyses of organics in dry soils, landfills, material from chemical waste dumps, and related matter. In the method a sediment sample is partially dried and extracted in an ultrasonic homogenizer with a 1:1 mixture of acetone-hexane. The residue is filtered and the extract concentrated in a Kuderna-Danish (K-D) apparatus. The concentrated extract may be analyzed directly by GC/MS or optionally separated into fractions with activated silica gel. The purpose of the acetone solvent is to provide a water-soluble wetting agent to stimulate desorption of organics from the solid material in the sample. The advantage of this method is that it uses a reasonably well tested extraction procedure and a familiar, well established optional preliminary fractionation scheme. The detection limit is not known, but is estimated at about 1–10 ug per kg of dry solids. Quality control must include a single reagent blank for each group of samples extracted on a given day. A reagent blank is the result of processing through the entire method all the same quantities of materials and labware, except that the sediment itself is not included. The result of the reagent blank analysis documents the background and reagent contamination.

It is not possible to determine recoveries with this method since it is not possible to spike a sediment and simulate natural adsorption conditions. One method of estimating the degree of desorption is to re-extract the processed sediment with fresh solvent or a second solvent pair such as methylene chloride-acetone, and determine if additional amounts of the same compound are obtained.

Purify the reagents and solvents as described under low boiling solvent extractions in section 3.1. Standard laboratory glassware and equipment is required for this method. See the equipment section under low boiling solvent extractions in section 3.1. There are a number of ultrasonic tissue homogenizers available and these give good results with sediment samples. A stainless steel container is recommended for blending to preclude breakage by stones or other hard objects.

Decant and discard the water layer over the sediment, and mix the residue to obtain as homogeneous a sample as possible. Transfer the sample to a shallow pan to partially air dry for about three days at ambient temperature. It is a good idea to select and remove from the sample large debris such as stones, broken glass, and bottle caps. Drying time varies considerably depending on soil type and drying conditions. Sandy soil will be sufficiently dry when the surface starts to split, but there should be no completely dry spots. Moisture content will be 50–80% at this point.

Weigh 50 g of the partially dried sample into a 600 ml stainless steel beaker. Add 50 g of anhydrous sodium sulfate and mix well with a large spatula. The quantities of sediment and drying agent may be scaled down by a factor of 2–10 if the sediment is from a known highly polluted source. Immediately after weighing the sample, weigh approximately 5 g of the partially dried sediment into a tared crucible. Determine the percent solids by drying overnight at 110C. Allow the crucible to cool in a desiccator before weighing. Optionally one may determine the percent volatile solids by placing the oven dried sample into a muffle furnace and heating at 550C for one hour. Again the crucible should be cooled in a desiccator before weighing. The concentrations of organic compounds are normally expressed as the percent of the dry solid after the 110C treatment.

To the 50 g sample add 250 ml of 1:1 acetone-hexane, lower the homogenizer into the beaker until it is approximately one cm from the bottom. Tilt the probe slightly about 15 degrees from vertical to prevent the solution from swirling out of the chamber while blending. Blend at maximum speed for 30 seconds. Raise the probe out of the solution and rinse it with acetone, allowing the rinsings to drain into the beaker. Filter the homogenized mixture through a 7 cm or larger medium porosity fritted glass funnel into a K-D flask fitted with a 10 ml ampul. Rinse the beaker with acetone and wash the filter cake with the rinse acetone. Clean the probe between samples with successive 15 second runs in tap water, distilled water and acetone. Wipe off the residual solid material between cleaning runs.

If the optional silica gel separation is to be used, concentrate the extract to 1 ml; if no silica gel separation is planned, concentrate the extract as described under low boiling solvent extraction in section 3.1.

For the silica gel separation, pack activated silica gel into a glass chromatography column (2 cm I.D.) to a height of 10 cm. To the 1 ml of extract in a small beaker add sufficient silica gel to adsorb the sample, and then evaporate the residual solvent with gentle stirring in a hood. Wet the column with about 20 ml of hexane and add the sample when the last of the 20 ml reaches the surface of the adsorbent. Rinse the beaker with hexane and add the rinsings to the column. The beaker should be rinsed several times with hexane and each succeeding eluent when the eluent is added to the column. Collect the fractions in K-D ampuls and elute the aliphatic fraction with 85 ml of hexane, the aromatic fraction with 85 ml of benzene or the safer toluene, and the polar fraction with 1:1 methanol-methylene chloride. Concentrate each fraction to 0.1–1 ml in the K-D apparatus, and proceed with the GC/MS analysis of each.

The GC columns and conditions recommended are the same as those used for low boiling solvent extraction. This procedure should be consulted for this information. Also see section 3.1 for sample GC/MS system dialogue.

3.4 FATTY TISSUE SAMPLES

This method is intended for the extraction of a broad spectrum of environmentally significant organics from fatty tissue, and the separation of these organics from the natural fats and oils prior to GC/MS analysis. The method has been tested mainly with fish tissue, but should be applicable to other types of tissues. The analysis of fish tissue is used to describe the method, with optional sample handling for the analysis of the whole fish, or just the potentially edible portions. The advantage of the method is that it is based on a significant amount of experience of several investigators over a number of years. The detection limit for the method is not known, but is estimated at 1–10 ug per kg of fish tissue.

The analysis of a reagent blank is required for each group of fish specimens extracted on a given day. The reagent blank analysis provides information about background and solvent contamination. Recovery data are not available for the total procedure since it is not possible to spike fatty tissue with known concentrations of organics and simulate natural conditions for the incorporation of these materials. Perhaps the best method of measuring the effectiveness of the extraction is to re-extract with a fresh portion of solvent, or with an alternative solvent, such as acetone-methylene chloride.

Solvents and other reagents should be purified as described under low boiling solvent extraction in section 3.1. For the analysis of large whole fish, a meat grinder is required. In all cases, a laboratory blender of about one quart capacity is also required. A gel permeation chromatograph equipped with a 2.5 x 50 cm column packed with BIO-RAD SX-2 beads is used to separate the natural fats and oils from the organics of interest.

Fish at the sampling site should be wrapped in aluminum foil, shipped in an ice chest packed with dry ice (preferred) or ice, and preserved in a freezer until analyzed. Small fish must be combined by sampling site and species to obtain the weight required for analysis. For the analysis of whole fish, the entire fish (or fishes) are ground directly, or, if necessary, chopped into pieces small enough to fit into the meat grinder. Grind the fish several times and thoroughly mix the ground material. Clean out any material remaining in the grinder and add this to the sample. For an analysis of the edible portions only, fillet the fish or fishes, and cut these into small pieces no larger than about two cubic centimeters each.

Add enough dry ice to the blender to completely cover the blades. Homogenize the dry ice for about 30 seconds, and then add about 25 g (weighed to the nearest 0.1g) of fish fillet chunks or ground whole fish along with more dry ice to the blender. Wait several minutes for the fish to freeze, homogenize for at least two minutes, or until the mixture is free of lumps, and then add about 75 g of purified anhydrous sodium sulfate to the blender. Homogenize for another two minutes, and pour the contents of the blender onto an aluminum foil sheet (one square foot). Homogenize additional dry ice and 25 g of anhydrous sodium sulfate and transfer this to the same aluminum foil. This operation serves to rinse the blender of residual fish residue. Mix the fish, sodium sulfate, and dry ice on the foil with a spatula. Carefully fold the aluminum foil into the form of an envelope, label, and place in a freezer at –15C for 8 – 12 hours. The dry ice will sublime and the residue should be in the form of a granular lump-free material.

Extract the mixture in a 250 ml Soxhlet extractor with 200 ml of a 1:1 mixture of acetone-hexane for 8 hours. Cool the extract, transfer it to a Kuderna-Danish (K-D) apparatus, and concentrate it to about 3 ml. Remove the last traces of solvent with a gentle stream of dry nitrogen, and weigh the resulting oil. Dilute the oil with methylene chloride to a concentration of 100 mg/ml.

Inject the total amount of diluted extract, in 5 ml aliquots, into the gel permeation chromatographic system. Elute each 5 ml aliquot with 225 ml of methylene chloride, at a flow rate of 3.5 ml per minute, and discard the first 160 ml of each eluate (this contains the natural oils and fats). Collect the balance (about 65 ml) of each eluate from each aliquot in the same 500 ml K-D flask. Concentrate the combined eluate containing the organics of interest to 5 ml. Equilibrate the gel permeation system with a 1:1 mixture of cyclohexane-methylene chloride. Inject the 5 ml of concentrated combined eluate and elute with 300 ml of 1:1 cyclohexane-methylene chloride at 3.5 ml per minute. Again discard the first 160 ml which contains the residual lipid material. The remaining eluate may be divided into 3–10 fractions or collected as a single fraction. Concentrate each fraction to 0.1 – 1 ml in a K-D apparatus as described under low boiling solvent extraction in section 3.1. The GC columns and GC/MS conditions used for the low boiling solvent extraction procedure are recommended for this method.

3.5 CHEMICAL DERIVATIZATION OPTIONS

Chemical derivatization methods have been developed extensively since 1955. Much of this effort was directed toward functional group selective reagents for derivatization of specific target compounds or narrow classes of compounds, e. g., steroidal alcohols. In terms of the broad spectrum approach defined in chapter 1 of this manual, a few reasonably general reagents, such as diazomethane, have been employed for derivatization of several broad classes of compounds, e. g., phenols, fatty acids, etc., that generally appear in the acid fraction of a low boiling solvent extract. Historically, derivatization procedures were needed to allow the gas chromatographic separation of compounds that were difficult or impossible to handle with existing liquid phases and solid supports. However, there has been a steady improvement in packing materials and open tubular column technology, and several classes of compounds, such as the steroids, phenols, and fatty acids, may now be conveniently chromatographed without derivatization. Under these circumstances, serious consideration must be given to the derivatization option, and the need for this additional step in any broad spectrum organic analysis. The advantages and disadvantages of derivatization are summarized in this section.

A clear advantage of chemical derivatization is that several or many different classes of compounds may be readily chromatographed with a few general purpose columns. Without derivatization, it may be necessary to change columns frequently to accomodate certain classes of compounds, phenols and fatty acids. Another advantage of derivatization is that atoms may be introduced into various compounds that give the compounds special properties that aid in their identification. Examples of this include the incorporation of a high fraction of heavy atoms such as deuterium or carbon-13, and naturally occurring isotope distributions, such as bromine-79 and bromine-81.

The disadvantages of derivatization include the additional cost, mainly in time and reagents, of these procedures. On the surface, the additional costs may appear slight, but additional quality controls will be required, and these may double or triple the number of GC injections which are time consuming and produce significantly more data that needs to be stored and interpreted. The real question is whether the additional cost is worth the additional information that may be acquired. Another disadvantage is the uncertainty about yield in any derivatization procedure. Unless adequate control samples are derivatized simultaneously under identical conditions with identical reagents, the measurement accuracy of the method may be seriously affected. Independent of the use of quality control samples, derivatization for broad spectrum organics analysis creates considerable uncertainty with regard to what the reagents are doing to many other components of the mixture. By-products and side reactions are not uncommon, and commercial reagents are frequently impure and introduce trace impurities that could be mistaken for sample components. The problems of reagent shelf life and the frequent poisonous and explosive nature of many reagents, e.g., diazomethane, are disadvantages that need to be carefully weighed in the cost-benefit equation.

In conclusion, it is recommended that for most broad spectrum analyses derivatization is not worth the effort, and that column switching is preferable. However, if certain target compounds, e. g., 2,4,6-trichlorophenoxyacetic acid, must be included in the analysis, derivatization is the only practical choice at this time. In the future, liquid chromatography – mass spectrometry interfaces may be developed to an extent to permit direct analysis of some classes of compounds not readily chromatographed without derivatization.

If it is necessary to form a derivative of a specific target compound, the commercially available reagents and procedures are strongly recommended. The Pierce Chemical Company of Rockford, Illinois, has been particularly active in developing these reagents, and a detailed discussion of the merits of various materials and procedures is beyond the scope of this manual. The choice of a reagent and procedure depends on the target compound or compounds, and usually several reagents and procedures will be available. Some important considerations in selecting procedures are the speed of the reaction, the convenience of the procedure, side reactions, and reproducibility. Often very pure and dry reagents are required, including generally anhydrous conditions. The commercially available reaction vials with Teflon lined septum tops are strongly recommended.

Finally, because of its particularly hazardous nature, a word about diazomethane is appropriate. It is a highly poisonous and very explosive yellow gas. Extreme caution must be exercised to avoid inhaling it or allowing it too near an open flame. The diazotization reaction must be conducted in an effective fume hood and safety glasses and shields are mandatory. No chipped, etched, or cracked glassware can be used because rough glass surfaces enhance the decomposition of diazomethane.

CHAPTER 4
DATA OUTPUT

The purpose of this chapter is to provide the information that is needed to output data with a computerized gas chromatography-mass spectrometer system. The chapter describes in detail the operation of programs that run on a Digital Equipment Corporation Model PDP-8 Computer.

One of the most fundamental and important programs that the user of a system requires is one that types out a list of programs and data files. All computer acquired data and all computer programs are stored as named files on either the disk or tape storage device. The names of the data files currently existing on the disk are required before output can begin. Only the files on the disk can be directly output when a disk operating system is in use. Output directly from tape is possible only with a tape operating system. If a disk operating system is in use, and output is required from magnetic tape, the file must first be copied from tape to disk before output can begin. (See the copy program in Chapter 7).

There are a number of programs available to list the programs and data files on a disk or tape. Each or these programs is somewhat different in hardware requirements, and user input. Only three of these programs are described in this section, and only three are supported under the revision E0 operating system. The program names and hardware requirements are as follows:

Program Name	Status	Hardware
LIST	Unsupported	Disk unit 0 and TC08 Dectape unit 1
LIST	Supported	Disk unit 0 and TD8E Dectape unit 0
MAGLST	Supported	9-track industry standard magnetic tape

Program Name	Status	Hardware
BCLIST	Unsupported	Disk units 0 or 1 and TC08 Dectape units 0–7
LSTDSK	Supported	Disk units 0–3 and TC08 Dectape units 0–7
LSTBLK	Unsupported	Disk unit 0 only
LSTDAT	Unsupported	Disk unit 0 only

Sample dialogue for the supported LIST program is as follows:

```
SELECT MODE:  LIST
TAPE?:  N

NAME    BLOCK

INIT$$     15
EXEC       17
EXMODI     47
CUEXEC     76
ETC.
```

In the dialogue a Y to the TAPE prompt causes the files on TD8E Dectape unit 0 to be listed. An entry of LIST to the system prompt and M to the TAPE prompt calls the program MAGLST which causes the files on the industry standard tape to be listed.

For users with TC08 Dectapes or users who wish to list the files on disk units 1–3, the program LSTDSK replaces the TC08 LIST program and BCLIST. LSTDSK requires 8 K of core memory and an extended arithmetic unit. The only input required is the disk number D0, D1, D2, or D3 or the TC08 Dectape number T0 – T7.

4.1 THE OUTP/CRT SOFTWARE

The plotter and printer programs that are called by entering OUTP in response to the system prompt are the original output programs for the PDP-8 GC/MS datasystem. As such, they are relatively simple programs that require only 4K of memory and no special PDP-8 hardware.

Nevertheless, they have some capabilities not available in the output software described under sections 4.2 and 4.3. Also, they are programs that are easy to use by the relatively inexperienced user. For these reasons they have been retained on the system, and are used extensively in many laboratories. Eight files are required on the system disk to run these programs: EXMOD1, PLOUT, PLOT, PLTP2, CUEXEC, RGC, TYPE, and BACKGD. These programs handle output from disk unit 0 only. Another program, PERCAL, may be used to retrieve stored operating parameters from any stored data file. This program is explained in more detail in section 2.7.

The CRT program requires 8K of memory and the files CRT, CRTRGC, CRTSP, CBCKGD, and DSPLAY must be present.

THE TOTAL ION CURRENT PROFILE

The total ion current profile (TICP) is defined as a normalized plot of the sum of the ion abundance measurements in each member of a series of mass spectra as a function of the serially indexed spectrum number. Data for this plot are acquired by continuous, repetitive acquisition of mass spectra as sample components emerge from a gas chromatograph, leak from a batch inlet system, or volatilize from a direct inlet system. Each point on the ordinate is the normalized sum of all the ion abundance data in a single mass spectrum; and, each point on the abscissa represents the spectrum number or a corresponding unit of time.

This same plot is referred to as a reconstructed gas chromatogram (RGC) but this nomenclature is not preferred as it does not accurately define a TICP. An RGC could just as well be the output of a flame ionization detector, as redrawn by a draftsman.

Cathode Ray Tube Terminal. A TICP may be output on a cathode ray tube (CRT) in a few seconds or on a pen and ink plotter in several minutes. Sample dialogue for the CRT output is as follows:

```
SELECT MODE: CRT
FILE NAME: SAMP1
DOUBLE DISPLAY?: N
TOTAL ION CURRENT PROFILE?: Y
EXTRACTED I C PROFILE?: N
```

If the user responds Y to the DOUBLE DISPLAY prompt before the carriage return, a TICP and a mass spectrum may be displayed together. Mass spectra outputs are discussed later. A response of Y to the EXTRACTED I C PROFILE?: prompt leads to dialogue which is used to display an extracted ion current profile (EICP). This option is described in a subsequent section.

Unless a TICP consists of only a few spectra, e.g., 25, it cannot be compressed on a small CRT and still permit the viewer to accurately select spectrum numbers for other operations. Therefore, commands exist to replot a TICP with a better view of a smaller area. These commands, which are entered from the keyboard after display of the TICP, are as follows:

E	further dialogue requests the extreme left hand spectrum number
C	entered repetitively after E for subsequent plots of narrower ranges of spectrum numbers beginning with the designated left hand value
R	restore last plot before expansion
G	the complete TICP is replotted
F	further dialogue requests filename for a different TICP

CTRL/L returns to the system prompt. Several other commands are available from the CRT keyboard during viewing of these plots. These produce EICP and mass spectra plots and are discussed in subsequent sections.

Plotter. The program for output of a TICP on the pen & ink plotter is more flexible than the CRT program. The output dialogue is as follows:

```
SELECT MODE:  OUTP
TOTAL ION CURRENT PROFILE:  Y
FILE NAME:  SAMP1
EXPAND BY:  press return
```

However, in this case the spectrum number resolution is fixed and quite adequate for selecting spectrum numbers for other operations. A response of a number between 1-100 to the EXPAND BY prompt, causes the total ion current data to be multiplied by the expansion factor before plotting. This will cause some peaks to be plotted with flat tops indicating the numerical data exceeds the range of the plot axis. Baseline as well as peak values are multiplied by the expansion factor. In general factors greater than about five are not very useful.

Long TICP plots from GC runs are best displayed immediately on the pen and ink plotter. The time requirement of several minutes is not that long, the quality plot will be needed ultimately for the record, and considerable time will be expended replotting for better viewing with the CRT. On the other hand the CRT is excellent for fast preview of TICP plots and the rapid examination of data from a single or several closely spaced peaks.

MASS SPECTRA HISTOGRAMS

Once a TICP is available, mass spectra are selected from individual peaks and these are used to identify the compounds causing the peaks. As with the TICP, mass spectra may be displayed on the CRT or the plotter. Since a TICP often consists of hundreds of mass spectra, and since it is often necessary to plot several mass spectra per peak, the fast CRT is of enormous value in previewing mass spectra. Ultimately, selected spectra may be plotted on the pen and ink plotter to make quality copy for reproduction in reports or for preservation in the permanent record.

Cathode Ray Tube Terminal. The dialogue for CRT output of mass spectra is as follows:

```
SELECT MODE:  CRT
FILE NAME:  SAMPLE
DOUBLE DISPLAY?:  N
TOTAL ION CURRENT PROFILE?:  N
SPECTRUM NUMBER:  3
SUBTRACT BACKGROUND?:  Y
SPECTRUM NUMBER:  2
```

In the CRT mass spectrum plot dialogue, a negative response to a TICP causes the program to assume that a mass spectrum is to be plotted. Normally the best spectrum number is found on the leading edge of a peak, about two-thirds of the way to the peak top. For a single component peak, the spectra will not differ much across the peak except for abundances at the very high and low amu values. For a multicomponent peak, several spectra can be used to determine which ions in a spectrum belong to a single component. Background subtraction is extremely important to remove extraneous ion abundance data from the spectrum of a compound. In addition to unresolved components, these extraneous ions are generated from continuous column bleed and background materials in the spectrometer that are not pumped out rapidly. Spectrum selection and background subtraction is often an iterative process, and several trials may be required to produce the optimum spectrum.

An affirmative response to the CRT double display prompt allows simultaneous display of two plots. The following dialogue illustrates the use of this feature to display a spectrum with background subtracted on the upper plot and the background spectrum on the lower plot.

```
SELECT MODE:  CRT
FILE NAME:  SAMPLE
DOUBLE DISPLAY?:  Y

UPPER GRAPH PARAMETERS:
TOTAL ION CURRENT PROFILE?:  N
SPECTRUM NUMBER:  3
SUBTRACT BACKGROUND?:  Y
SPECTRUM NUMBER:  2

LOWER GRAPH PARAMETERS:
SPECTRUM NUMBER:  2
SUBTRACT BACKGROUND?:  N
```

The CRT commands E, C, R, G, F, and CTRL/L are also available after mass spectral plotting. In this case E sets the extreme left hand amu value. Another command, S, may be entered after viewing a TICP or a mass spectrum. It requests a spectrum number for plotting as well as information for background subtraction.

Plotter. The program for plotting a mass spectrum on the pen and ink plotter is more flexible than the CRT program and also permits one to print digital mass spectral data on the console printer or CRT as well as plot on the plotter. The output dialogue used to plot a mass spectrum is:

```
SELECT MODE:  OUTP
TOTAL ION CURRENT PROFILE:  N
EXTRACTED ION CURRENT PROFILE:  N
PLOT SPECTRUM?  Y
FILE NAME:  REF2
SPECTRUM NUMBER:  49
AMPLITUDE EXPANSION?:  press return
MINIMUM VALUE %:  1
SUBTRACT BACKGROUND?:  Y
SPECTRUM NUMBER:  44
BACKGROUND AMPLIFICATION:  press return
SAVE SUBTRACTED FILE?:  N
NORMALIZE ON:  press return
```

A plot of the reference compound mass spectrum using the dialogue is shown in Figure 4.1. To illustrate several of the plot options the following dialogue was used to generate the spectrum in Figure 4.2.

```
SELECT MODE:  OUTP
TOTAL ION CURRENT PROFILE:  N
EXTRACTED ION CURRENT PROFILE:  N
PLOT SPECTRUM?:  Y
FILE NAME:  REF2
SPECTRUM NUMBER:  49
AMPLITUDE EXPANSION?:  Y
THRESHOLD %:  5
EXPAND BY:  5
MINIMUM VALUE %:  2
SUBTRACT BACKGROUND?:  Y
SPECTRUM NUMBER:  44
BACKGROUND AMPLIFICATION:  10
SAVE SUBTRACTED FILE?  Y
FILE NAME:  REF2BS
NORMALIZE ON:  press return
```

An affirmative response to the AMPLITUDE EXPANSION? prompt leads to dialogue in which a per cent relative abundance threshold and an expansion factor are entered by the user. The product of the threshold and expansion factor may not exceed 30. The principal application of this option is to emphasize very weak ion abundances in the spectrum plot. The minimum value option is used to eliminate plotting of very weak abundances below the minimum. Normally the minimum value and amplitude expansion options are not used in the same plot. Background amplification may be used to enhance the ion abundance data in the background spectrum before it is subtracted from the sample spectrum. If the user chooses to save the background subtracted spectrum, a file name must be supplied for the data. All spectra are automatically plotted with the most abundant ion set to 100% relative abundance. However, a user may enter an alternative mass number in response to the NORMALIZE ON response, and the abundance of that mass will be set to 100% with a proportionate adjustment for all other abundances. This has the effect of amplifying weak ion abundances without the scale expansion shown in Figure 4.2

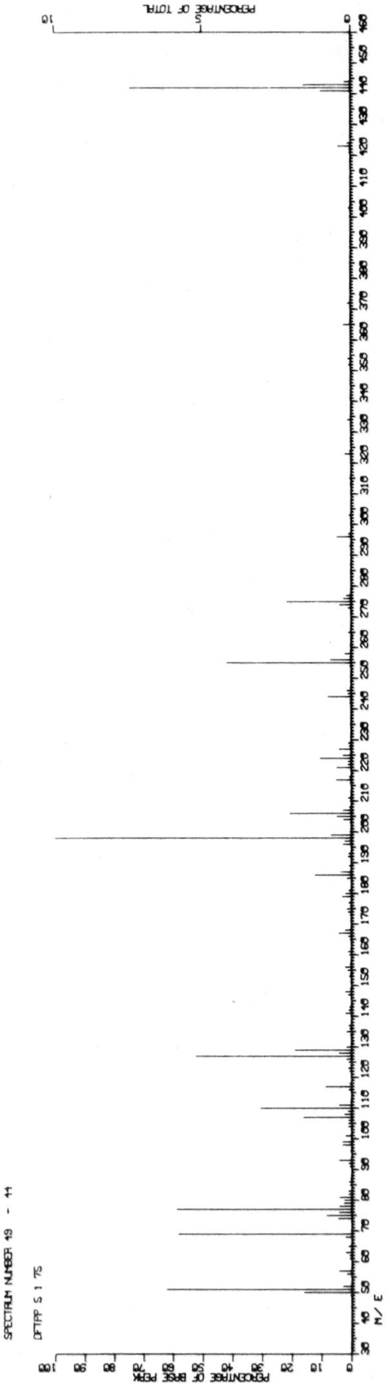

Figure 4.1 The mass spectrum of a GC/MS reference compound, decafluorotriphenylphosphine.

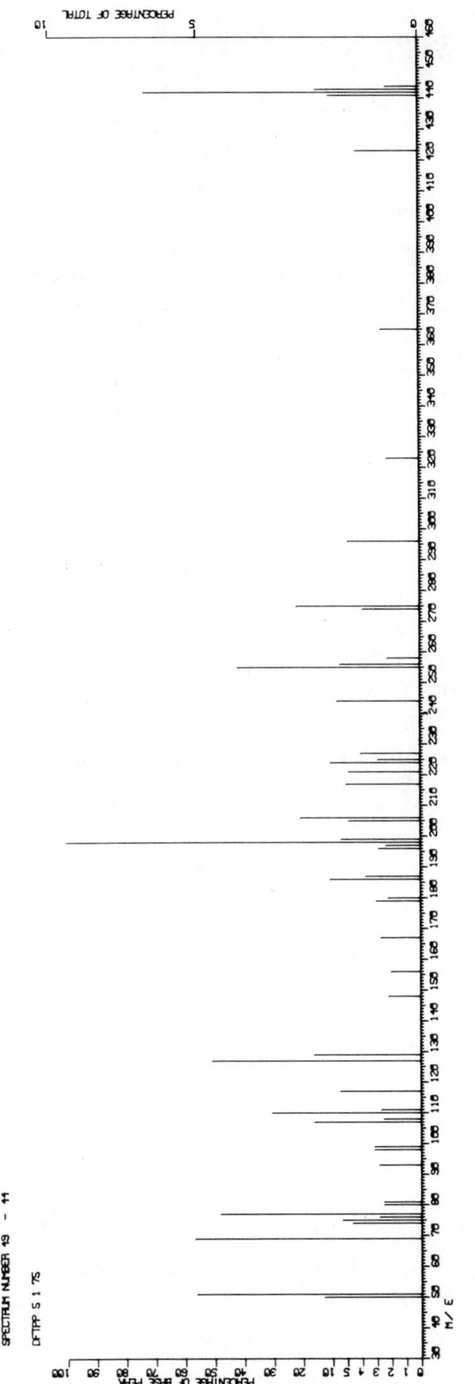

Figure 4.2 The mass spectrum of a GC/MS reference compound, decafluorotriphenylphosphine, with plot options.

PRINTED MASS SPECTRAL DATA

Printed mass spectral data may be produced on either the CRT or the console printer, but the OUTP mode is required in both cases. Output is directed to one or the other via the switch on the back of the CRT. There are two possible responses to the PRINT SPECTRUM?: prompt. A Y response will cause mass numbers, per cent relative abundances, and per cent of total abundances to be printed. An A response will cause mass numbers and the corresponding absolute abundance data to be printed. Absolute abundances are in analog-to-digital converter units. Standard printout dialogue allows the user some of the same options as the plotter routine:

```
SELECT MODE:  OUTP
TOTAL ION CURRENT PROFILE:  N
EXTRACTED ION CURRENT PROFILE?:  N
PLOT SPECTRUM?:  N
PRINT SPECTRUM?:  Y
FILE NAME:  SAMP2
SPECTRUM NUMBER:  42
PARTITIONED OUTPUT?:  N
MINIMUM VALUE %:  5
SUBTRACT BACKGROUND?:  Y
SPECTRUM NUMBER:  40
BACKGROUND AMPLIFICATION:  press return
SAVE SUBTRACTED FILE?:  N
NORMALIZE ON:  press return
```

AMU	INTENSITY	PERCENT OF TOTAL INTENSITY
51.0	9.27	4.45
63.0	7.44	3.58
64.0	8.12	3.90
102.0	9.21	4.43
102.0	9.21	4.43
126.0	7.39	3.55
127.0	13.90	6.68
128.0	100.00	48.09
129.0	11.45	5.51

However, in place of the plotter AMPLITUDE EXPANSION? prompt, the PARTITIONED OUTPUT? prompt is printed. A Y response permits the output of selected ion abundance data. The following application of partitioned output shows data from a spectrum of perfluorotri–n–butyl amine (PFTBA). This spectrum was measured immediately after calibration while PFTBA was in the spectrometer. The partitioned output gives a rough idea of the condition of the spectrometer.

```
SELECT MODE:  OUTP
TOTAL ION CURRENT PROFILE:  N
EXTRACTED ION CURRENT PROFILE?:  N
PLOT SPECTRUM?:  N
PRINT SPECTRUM?:  Y
FILE NAME:  1
SPECTRUM NUMBER:  1
PARTITIONED OUTPUT?:  Y
MASS RANGE:  67–72; 219–222; 500–505; 612–616
MINIMUM VALUE %:  press return
SUBTRACT BACKGROUND?:  N
NORMALIZE ON:  press return
```

AMU	INTENSITY	PERCENT OF TOTAL INTENSITY
67.0	.04	.02
68.0	.49	.22
69.0	100.00	46.06
70.0	1.38	.63
71.0	.02	.00
72.0	.00	.00
219.0	25.23	11.62
220.0	1.09	.50
221.0	.02	.00
500.0	.00	.00
501.0	.00	.00
502.0	.84	.38
503.0	.09	.04
613.0	.00	.00
614.0	.15	.07
615.0	.02	.00
616.0	.00	.00

Partioned output of digital data is analogous to the extracted ion current profile that is discussed in the next section.

THE EXTRACTED ION CURRENT PROFILE

The extracted ion current profile (EICP) is defined as a plot of the change in relative abundance of one or several ions as a function of the serially indexed spectrum number. It is called an extracted ion current profile because the data are literally extracted from the total ion current profile (TICP) data in a non-real time process. The EICP is therefore analogous to partitioned output of digital data. However, it should not be confused with real-time selected ion monitoring (SIM) and the corresponding selected ion current profile (SICP) that are discussed in Chapter 6. The SIM technique is a real time process in which ion abundance data are measured at only selected masses as components emerge from the gas chromatograph or other inlet systems. The SIM technique produces a real increase in signal/noise by time averaging random noise. The EICP produces an apparent increase in sensitivity by removing from the TICP the ion abundance data from background, unresolved components, and other irrelevant ions. The terms limited mass output or limited mass search are often used, but are less meaningful than extracted ion current profile to describe this plot.

On the CRT, the EICP is generated by an affirmative response to the TOTAL ION CURRENT PROFILE?: prompt and an affirmative response to the EXTRACTED I C PROFILE?: prompt. The user is then requested to enter a mass or mass range. The CRT commands E, C, R, G, and F are available as described above, and an L command may be used to change the mass range in the EICP plot. The dialogue for a CRT EICP is as follows.

```
SELECT MODE:  CRT
FILE NAME:  SAMPLE
DOUBLE DISPLAY?:  N
TOTAL ION CURRENT PROFILE?:  Y
EXTRACTED I C PROFILE?:  Y
MASS RANGE:  149
```

In response to the MASS RANGE? prompt, the user may insert up to eight individual masses or mass ranges separated by semicolons. The EICP produced is a sum of the abundance data for each of the masses or mass ranges specified. Figure 4.3 shows a TICP and the EICP for mass 149 from the same data file. The compounds that produce mass 149 ions are clearly highlighted in the EICP. The plotter output dialogue is similar to the CRT dialogue.

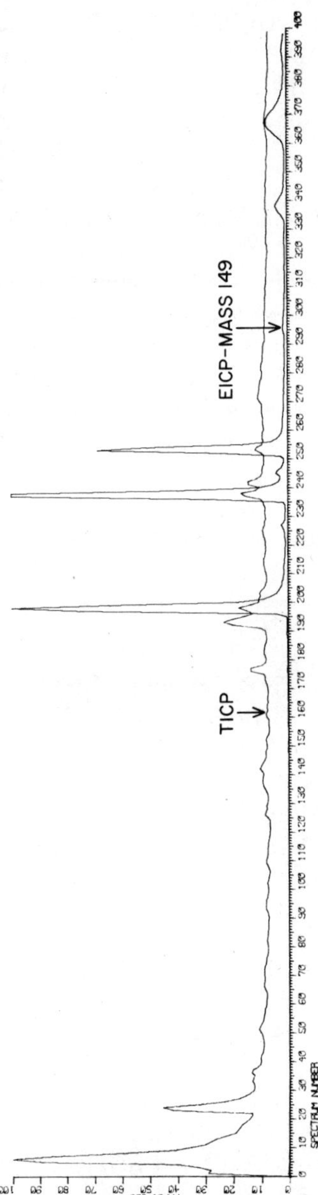

Figure 4.3 The extracted ion current profile for mass 149 and the corresponding total ion current profile.

```
SELECT MODE:  OUTP
TOTAL ION CURRENT PROFILE:  N
EXTRACTED ION CURRENT PROFILE?:  Y
FILE NAME:  SAMP1
MASS RANGE:  149
EXPAND BY:  press return
```

QUEUED OUTPUT

The user may defer actual output of various pen and ink plots and printouts until a series of output commands have been stored for later execution. In response to the system prompt, the user enters the command QUEUE. The system immediately returns to the system prompt. The user enters the information required to generate an output. However, at the completion of the dialogue the plot is not executed, but the system returns to the system prompt. A series of output dialogues may be completed by a repitition of this procedure. The queue is ended with the command ENDQ in response to the system prompt. All commands are then sequentially executed. The user must delete the $ $ QUE file before another queued output can begin. The following dialogue is an example of a queue for two spectrum plots.

```
SELECT MODE:  QUEUE
SELECT MODE:  OUTP
TOTAL ION CURRENT PROFILE:  N
EXTRACTED ION CURRENT PROFILE:  N
PLOT SPECTRUM?:  Y
FILE NAME:  803D
SPECTRUM NUMBER:  34
AMPLITUDE EXPANSION?:  press return
MINIMUM VALUE %:  1
SUBTRACT BACKGROUND?:  Y
SPECTRUM NUMBER:  31
BACKGROUND AMPLIFICATION:  press return
SAVE SUBTRACTED FILE?:  N
NORMALIZE ON:  press return

SELECT MODE:  OUTP
TOTAL ION CURRENT PROFILE:  N
EXTRACTED ION CURRENT PROFILE:  N
FILE NAME:  816D
SPECTRUM NUMBER:  67
AMPLITUDE EXPANSION?:  press return
```

```
MINIMUM VALUE %:  1
SUBTRACT BACKGROUND?:  Y
SPECTRUM NUMBER:  60
BACKGROUND AMPLIFICATION:  press return
SAVE SUBTRACTED FILE?:  N
NORMALIZE ON:  press return

SELECT MODE:  ENDQ
```

4.2 THE MSSOUT OUTPUT SOFTWARE

This output system has many of the same basic capabilities as the output and CRT software described in section 4.1. However, the MSSOUT software has flexibility and capabilities not found in the standard output programs. This progam requires the presence of an extended arithmetic element (EAE) in the PDP-8 computer. The EAE unit consists of two printed circuit boards labeled M8340 and M8341. These provide enhanced processing speed for certain types of computer instructions.

The MSSOUT software also requires the presence of seven files on the system disk: MSSOUT, MOUTP1, MOUTP2, MOUTP4, MOUTP5, MOUTP6, and INCCHR. Another file, $S1$, is generated by the program. As with many other choices, there are advantages and disadvantages to adopting the MSSOUT software. The principal disadvantage, in addition to the hardware capital investment, is that a set of user commands need to be mastered in order to use the capability. The MSSOUT program is non-prompting (non-interactive). The user must know what commands to enter and their proper sequence. Another disadvantage is that there is a bug in the MOUTP4 file that has proven difficult to locate and repair. This bug causes flat topped peaks that are really not saturated. Another version of MOUTP4 exists in which the flat topped peak bug was repaired, but spectra are sometimes missed. A temporary fix is to retain both versions on the system disk under other names. This allows the user to delete the current MOUTP4 and rename with the COPY program one of the versions. The choice depends on which bug the user wants to live with. The file MTP4/1 is a version of MOUTP4 that gives flat topped peaks; the file MTP4/2 gives missing spectra.

The advantages of the program appear to be very significant and include the following:

1. The user has the capability to designate within limits the size of various plotter displays and alphanumeric labels.
2. The user has the ability to switch rapidly between CRT and plotter displays within the same program.
3. Because of the EAE hardware, processing speed is improved somewhat compared to the standard software.
4. Mass spectra histograms may be plotted over a smaller mass range than was actually observed during the original data acquisition. Therefore if the 33–450 amu range was observed, but no ions were observed above mass 300, the histogram plot may be limited to 33–300 amu.
5. Similiar capabilities exist for Total Ion Current Profile (TICP) and Extracted Ion Current Profile (EICP) plots, i.e., portions of chromatograms that contain the peaks may be plotted, and the leading or trailing sections that contain no peaks may be deleted from the graphics display.

The MSSOUT outputs all begin with the same basic dialogue:

```
SELECT MODE:  MSSOUT

RUN NAME:  STD
NBS STD 80NG EACH

*
```

In this dialogue the run name means the same as the filename in the standard output software. After the run name is entered, the program immediately prints the title attached to that file. If an informative title was entered, this will confirm the identity of the data file. After this an asterisk is printed. This is a signal that the program is ready to receive commands from the user. If the user wishes to change the data file (run name) this may be accomplished anytime after an asterisk is printed. Simply enter the NA (name) command and the program returns to the RUN NAME: prompt.

```
*NA

RUN NAME:  213
NBS SAMPLE 213 5UL
```

Anytime after an asterisk is printed the user may enter the EX (exit) command which causes the program to return to the system prompt. The user commands for various outputs are described in the following sections. These are organized similarly to section 4.1 with sections on the TICP, Mass Spectra Histograms, and Printed Mass Spectral Data, the EICP, and Queued Output.

THE TOTAL ION CURRENT PROFILE

The sequence of commands in the sample dialogue will cause a TICP to be plotted on the plotter. The TICP will begin at spectrum number 20 and end at spectrum number 220. The plot will be 3 inches high by 5 inches long. Figure 4.4 is the result of these commands:

```
*CM
ENTER MASSES TO COLLECT. MAX= 24


*EM
ENTER MASSES TO PLOT
T

*ST=20

*EN=220

*HE=3

*LE=5

*PP
SCAN
```

This sample dialogue illustrates the use of some of the commands that may be used to generate a TICP. The CM (collect masses) command is mandatory and causes the program to gather and save in core memory the data necessary to plot a TICP and, optionally, a number of EICP's. If the user responds with a RETURN to the ENTER MASSES TO COLLECT prompt, as was done in the example, only data for a TICP will be collected, and it will not be possible to display an EICP. However the user may reenter the CM anytime an asterisk is printed, and cause the program to collect a new set of data. The MAX=X gives the maximum number of EICP's that may be saved in core memory, and this is explained further in the section on the EICP. If MAX=0, the data for a TICP is saved after entering RETURN. If specific masses are entered, the data necessary for a TICP is always gathered in addition to the EICP data.

The EM command is optional in the most recent version of MSSOUT, and this revision should be placed on all disks. The functions of the EM command are explained in the section on the EICP.

The ST (start) = X (where X = an integer spectrum number) and EN (end) = X commands are optional, and allow the user to define starting and ending spectrum numbers for the TICP. If these commands are omitted, the default values are the first and last spectrum numbers in the file.

The HE (height) = X (where X= integer inches) and LE (length) = X commands are also optional, and apply to the plotter only, not the CRT. They allow the user to define the height and length of the plot up to the maximum size of the paper. If these commands are omitted, the default size is 5x8 inches. The final command that must be entered is either a PP for a plot on the plotter or a PM for a display on the CRT.

An extremely important feature of MSSOUT is that the values entered for all commands are saved until they are changed or until the program is restarted. This feature results in a number of convenient possibilities for the user. For example one may issue a PP immediately after a PM and produce the same plots on the plotter and CRT. Alternatively, the user may change just one or two commands then enter the PP or PM, and immediately produce a modified plot without reentering all the commands. The command OP (options) will print out a list of all current command options, and this should be used in the event the user needs to recall the current option selections. The values of the options shown in the following list specify the default values, and these will be used for all plots until they are changed by the user. In some cases, e.g., HE, the actual default value is given; however in most cases, e.g., CM, a zero specifies the default situation.

```
*OP

CO      0
ST      0
EN      0
BA      0
MA      0      0
LI      0
HE      5
LE      8
TH      1
CS      0
LA     50
FR      Y      0
```

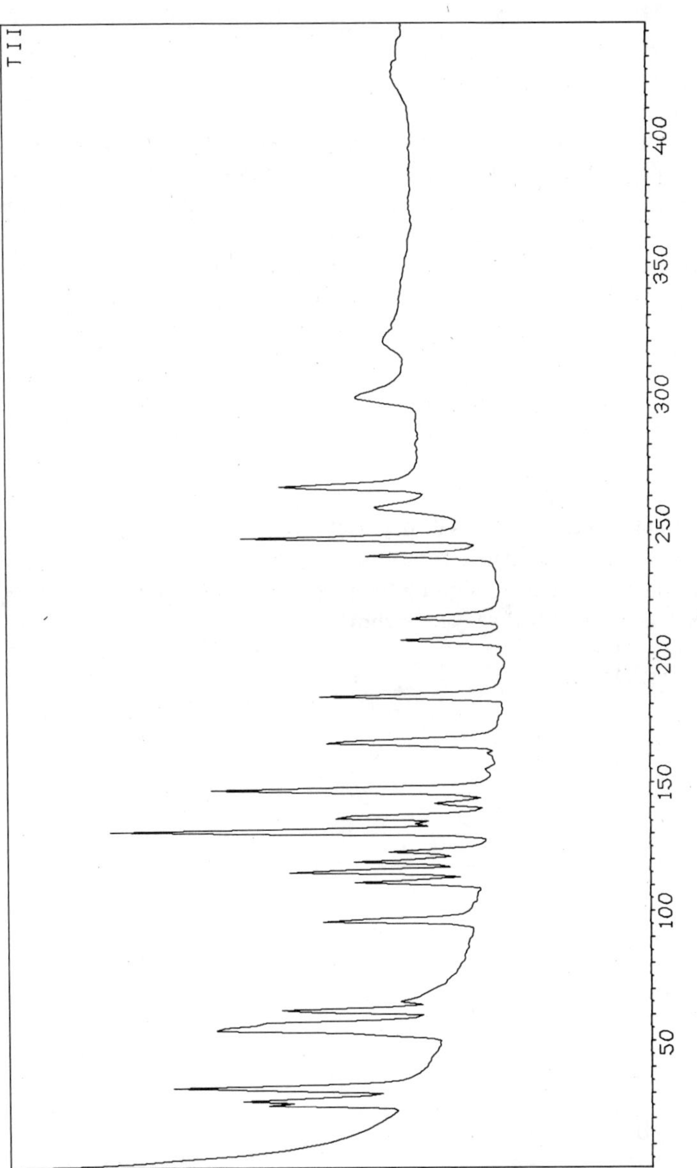

Figure 4.4 A total ion current profile generated with MSSOUT

CM	0	
SM	0	
MC	S	0
SU	0	
SF	100	

These option values are saved during a terminal session even though the user changes the data file with the NA command.

An IN (integrate) command also exists to integrate peak areas. However, this is discussed in section 6.2 along with other quantitative analysis methods. If the data was acquired with the RIB interface and time data was stored with the datafile, MSSOUT may be used to give plots with time in minutes on the x axis. Setting the parameter MC=T gives the time axis. The default value is MC=S or spectrum number.

MASS SPECTRA HISTOGRAMS

The sequence of commands in the sample dialogue will cause a mass spectrum to be plotted on the plotter. A background spectrum will be subtracted, and the plot will begin at mass 50 and end at mass 310. The filename (run name), spectrum number (scan number), and axis labels will be in one-quarter inch high characters. The size of the plot will be 6 x 8 inches. Figure 4.5 is the result of these commands:

```
*SC

SCAN NUMBERS:
216

*SU
ENTER BACKGROUND:   213

*ST=50

*EN=310

*CS=25

*HE=6

*LE=8

*PL
```

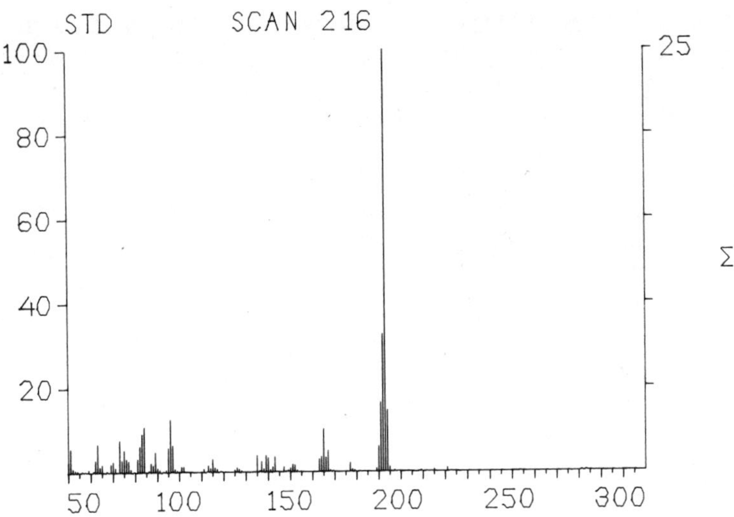

Figure 4.5 A mass spectrum plot generated with MSSOUT

The SC command is mandatory for a mass spectrum plot and the spectrum number is requested. Several spectra may be specified by a series of spectrum numbers separated by commas. The only other mandatory command is either a PL for a plot on the plotter or a DI for a display on the CRT. After a spectrum is displayed on the CRT, entry of a return will cause the CRT to be paged for the next spectrum, if more than one was specified, or the next command. If more than one spectrum is specified with the SC command, and output is to the plotter, the plots are queued automatically one after the other. There are a large number of optional commands that may be used to refine these graphics outputs. Some of these apply only to displays on the CRT, some apply only to plots on the plotter, and some apply to both outputs.

All of the following options apply to both the CRT and plotter. The SU command causes a background subtract prompt. If more than one background spectrum is specified, the background subtraction process appears to be aborted, but no error message is output by the program. The ST (start) = X (where X = an integer mass) command defines the starting mass of the plot, and the EN (end) = X command defines the ending mass of the plot. The default values for ST and EN are the first and last masses observed. The BA = X (where X = an integer mass) may be used to specify the mass used to normalize the mass spectrum plot. The default value is the base peak in the spectrum. The LA = X (where X = an integer mass) may be used to specify the mass interval between labeled tick marks on the mass axis of a spectrum plot. The default value is 50 amu. The SF = X (where X is a integer factor) may be used to specify the amount of amplification of the background spectrum before background subtraction. The default value is 100 which corresponds to an amplification factor of one. Therefore entry of SF = 150 will amplify the background file by 50%. Entry of SF = 50 will reduce the background spectrum by a factor of two. The MA = X,Y (where X = an integer mass number and Y = a magnification factor) may be used to specify that beginning at mass X all ion abundances are to be multiplied by the factor Y before plotting. The mass where magnification begins is indicated on the plot as in Figure 4.6. The capability to average several spectra, e.g., across a TICP peak, is another optional procedure which is accomplished with the dialogue:

```
*SC

SCAN NUMBERS:
A

*AV
ENTER SCANS TO AVERAGE: 214-220
*
```

The average spectrum plot is completed with the other standard commands as above.

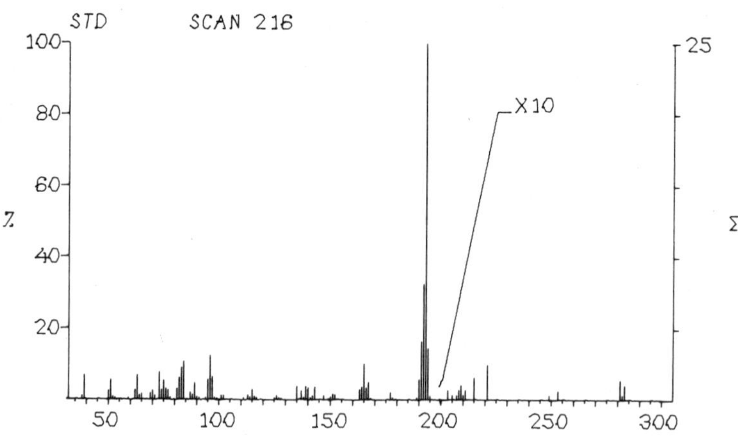

Figure 4.6 A mass spectrum plot generated with MSSOUT

A summary of optional commands that apply to the plotter and the CRT is as follows:

SU Subtract background
ST Starting mass
EN Ending mass
BA Normalize on
LA Interval of mass axis labels
SF Background amplification factor
MA Ion abundance multiplier
A/AV Average spectra

As previously described under the total ion current profile, values entered for most of these options are saved until changed or until the program is restarted. The current values or default values of the optional commands may be printed with the OP command. A command RE (repeat) is also available for mass spectrum plots. This causes a repeat of the plot that was most recently requested.

There are three commands that apply just to CRT displays. The DI command is the command which causes the actual CRT display and this was described above. The LI = X (where X = a low integer value, e.g., 1,2, or 3) is an option that defines the number of horizontal regions the CRT is divided into for an expanded presentation of a mass spectrum. The practical limitation of LI is probably 3 or 4, and the default value is selected by the program depending on the mass range scanned. Finally the CO = X (where X is an integer) command may be used to automatically make X number of hard copies of the CRT display.

There are several commands that apply just to the plotter. The PL command causes the plot to begin, and this was described above. The FR command causes a line (frame) to be drawn across the top of the mass spectral plot. The DF comand deletes this frame and is the default choice. The HE = X (where X = height in inches) and LE = X commands have the same meaning as described under the total ion current profile. These options may be used to specify the size of the plot within the limits of the paper size. The default size is 5 x 8 inches. Finally the CS = X (where X = an integer value) may be used to define the size of the alphanumeric characters used in the plot labels. A size of one inch is expressed by CS = 100, one half inch by CS = 50, and one-quarter inch by CS = 25, etc.

In summary, the commands that apply either to the CRT or the plotter, but not both are as follows:

DI Begin CRT plot
LI Division of CRT into regions
CO Automatic CRT copies

PL Begin plotter plot
FR Draw line on top of histogram
DF Delete line
HE Set height of plot
LE Set length of plot
CS Define character size

PRINTED MASS SPECTRAL DATA

Lists of digital data may be printed on the CRT or printer using the commands TA and RD. The TA command gives a normalized list and the RD command generates raw data in terms of analog to digital converter counts. A spectrum number or numbers must be specified first with the SC command.

THE EXTRACTED ION CURRENT PROFILE

The EICP plot is a variation on the total ion current plot, and all the commands that apply to the TICP also apply to the EICP. However the method of using the CM and EM commands is crucially important. The CM (collect masses) command is required as in the TICP, but the individual masses of interest must be specified in response to this command. Mass ranges are not accepted as a response to the CM command. The printed message that follows the CM command informs the user of the maximum number of individual masses that may be specified. This maximum will vary as a function of the size of the data file.

```
 *CM
ENTER MASSES COLLECT. MAX=   24
 143,167,193,208
```

After the collection is completed the user may display the plots in a number of sequences, and create other types of plots. These very flexible options are described in this section. The most basic option is to display the EICP's specified in exactly the order and quantity given in the CM command. This may be accomplished on the CRT or plotter, but the TICP is always placed on the bottom of the display. The EICP's are then placed in order, from the bottom to the top, as they were listed with the CM command. This list is saved until another CM command is issued, or the program is restarted.

The functions of the EM command are to redefine the order in which various plots are displayed on a single CRT or plotter display, and to generate subsets of the plots specified in the CM command. For example, if the EM command below were entered, the sequence of plots from the bottom to the top of the display would be a TICP, a mass 169 EICP, a sum plot (defined later), a TICP, a mass 197 EICP, a mass 213 EICP, and another TICP.

```
*EM
ENTER MASSES TO PLOT
T,169,S,T,197,213,T
*
```

Figure 4.7 is an example of this type of display using a different response to the EM command. The sequence of plots defined in the EM command is saved until another EM command is issued or the program is restarted, but the current value of EM does not appear in the options list. Mass ranges are not accepted by the EM command.

The EM command may also be used to define subsets of masses prior to the IN (integrate) command that is described in section 6.2.

The sum plot referenced above displays the sum of the abundances of two or more masses. The S response to EM is not functional until a sum is defined by the SM command as follows.

```
*CM
ENTER MASSES TO COLLECT. MAX= 24
143,167,193,200

*EM
ENTER MASSES TO PLOT
S

*SM
ENTER M/E'S TO SUM
143, 167,193

*PP
SCAN
```

This dialogue produces a single plot that is an EICP for the sum of the abundances of the three masses specified. There is one other command, the

BA = X (where X is an integer mass), that may be used to specify the abundance used to normalize any EICP plot.

In summary the commands that may be used for the EICP are as follows:

PP Causes actual plot on the plotter
PM Causes actual plot on the CRT
ST Defines starting spectrum number
EN Defines ending spectrum number
HE Defines height of plot on the plotter
LE Defines length of the plot on the plotter
SM Requests abundances to be summed
BA Normalize EICP on mass

The user should note that Figure 4.7 has a line drawn across the top of each EICP and that this may not be deleted with the DF command because this command does not apply to TICP and EICPs.

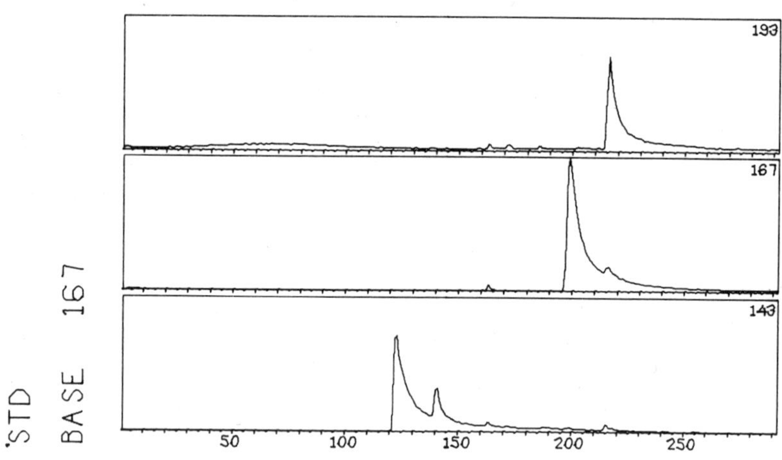

Figure 4.7 An extracted ion current profile generated with MSSOUT

QUEUED OUTPUT

Queued output of a series of separate displays with MSSOUT is only possible with mass spectra plots. The user should enter, in response to the message printed after the SC command, a series of spectrum numbers separated by commas. The SU command may be used to subtract the same background spectrum from each. After each spectrum is plotted on the CRT, pressing the return causes the program to immediately produce the next plot. However if a CO command is used, an automatic hard copy of each will be generated, and the next spectrum produced without the return. A similar series of plots may be queued to the plotter.

4.3 INTERACTIVE GRAPHICS ORIENTED OUTPUT SOFTWARE (IGOOS)

This output system has many of the same basic capabilities as the plotter and CRT software described in sections 4.1 and 4.2. However, the IGOOS has significant additional capabilities not available with the previously described output systems. This software requires the presence of an extended arithmetic element (EAE) in the PDP-8 computer. The EAE unit consists of two printed circuit boards labeled M8340 and M8341. These provide enhanced processing speed for certain types of computer instructions.

The standard TICP and EICP capabilities require the presence of three files on the system disk: BCLRGC, RGCOV1, and RGCOV2. Mass spectrum plotting requires the files CRTPLT, PLOVL1, and PLOVL2. As with many other choices, there are advantages and disadvantages to adopting the IGOOS. The principal disadvantage, in addition to the hardware capital investment, is that a set of user commands need to be mastered in order to use the capability. The IGOOS has its own set of commands and the user must know what commands to enter and their proper sequence. Another disadvantage is the presence of several program bugs. These are documented along with the user instructions. The advantages of the program appear to be very significant and include the following:

1. The user has the capability of selecting spectrum numbers from CRT graphics displays by placing the cross hair cursors over the appropriate place on the display, and entering a single character command at the keyboard.
2. The convenient spectrum number selecting capability described above may be used to build Q files consisting of spectrum numbers and corresponding background spectrum numbers. These Q files may be used to drive other programs that plot spectra on the CRT

or plotter, that abbreviate spectra for matching against a local or remote database with a spectrum matching program, or that perform quantitative analyses.

3. The same basic software may be used to generate CRT and plotter displays of various graphics, with the CRT used as a convenient preview of what will appear on the plotter.
4. Because of the EAE hardware, processing speed is improved somewhat compared to the software described in section 4.1.
5. Finally there is the capability to display TICP and EICP plots exactly above one another on the CRT or plotter. Up to six plots can be overlayed in this way.

TOTAL AND EXTRACTED ION CURRENT PROFILES

The user entry into this program is with the following dialogue which specifies the first TICP or EICP to be displayed on the CRT:

```
SELECT MODE:  BCLRGC
FILE?    CINTAP
Q-FILE NAME?   QCINTA
RANGE?
EXPAND BY?
```

In response to the FILE? prompt the user enters the data filename. A Q-filename must be entered even though the user has no intention of setting up a Q file. The program will perform if a RETURN is entered in response to the Q-FILE NAME? prompt, but an unnamed Q file will result if a Q file is initialized, and this will cause problems. If a Q file is not initialized and closed as explained below, the Q-filename will not be saved in the disk directory. In response to the RANGE? prompt the user may enter a RETURN for a TICP or a single mass or mass range for an EICP. In response to the EXPAND BY? prompt the user enters a RETURN for no expansion, or an integer number which is a multiplication factor for profiles containing only weak peaks. A CRTL/L during this or any subsequent dialogue causes a return to the system prompt.

At this point the designated TICP or EICP is displayed on the CRT screen, and the cross hair cursors are activated. If the data file consists of 260 or fewer mass spectra, the entire TICP or EICP will be displayed. If the file contains more than 260 spectra, only the first 260 points are displayed. A block of 260 spectra is referred to as a screen load. The TICP's and EICP's are displayed in 260 spectra screen loads in order to allow sufficient accuracy in positioning of the cross hair cursor on a given spectrum number.

With the first screen load displayed the user has only two options. Entry of a series of RETURNS will cause all sections of the TICP or EICP to be displayed in screen loads of 260 spectra with a 20 spectra overlap between each. Alternatively with any given screen load the user may initialize a Q-file by entry of the character Q. This will activate the S and B keys until Q is entered again which closes the Q-file.

With the S and B keys activated the user may save spectrum numbers in the Q-file by positioning the horizontal and vertical cross hairs over the desired point on the profile, and striking S or B. The S key is normally used to save spectrum numbers corresponding to the apexes of peaks, and the B key is used to save the corresponding background spectra. The B key cannot be used except after an S is entered. However, a succession of S's may be entered without saving any background spectra. The spectrum numbers saved are not required to correspond to apexes and background. For example, they could be used to save spectrum numbers that correspond to the limits of intergration for a quantitative analysis program. Entering a RETURN will cause the next screen load to be displayed while the Q file is still open. At the conclusion of the saving of all spectrum numbers, the Q key must be pressed to close the Q file and cause it to be saved in the disk directory.

After the last screen load of a given data file is displayed, certain single character commands are activated to permit other displays or plots. These commands are the O, RETURN, F, R, and H. Several of these commands return to the RANGE? prompt for specifications for displays, and several others cause the program to remember or forget previous specifications for CRT plots.

Entry of an O causes the program to return to the RANGE? prompt for specifications for another TICP or EICP that will be overlayed on the first plot on the CRT. Up to six plots may be overlayed on the CRT, and the program remembers which was specified first and most recently. There is a reported bug in the O option that may be present in other options. The bug appears only when a Q file is initialized during the viewing of a TICP. If the same TICP is overlaid with an EICP using the O option, the TICP is displayed incorrectly.

Entry of a RETURN causes the program to forget all specifications for previous displays and return to the RANGE? prompt.

Entry of an F causes the program to forget all specifications for previous displays except the first, and to display the first TICP or EICP specified on the CRT. This command has a slight bug in some versions of the program. It assumes the first display specified was a TICP; but, if it was an EICP, the data will be displayed correctly, but the EICP will be labeled a full mass range plot.

Entry of an R causes the program to forget all specifications for previous displays except the first, and to return to the RANGE? prompt for specifications for a TICP or EICP. This is then displayed on the CRT overlayed with the first plot specified.

Entry of an H causes the last TICP or EICP specified to be plotted on the plotter. Before this begins the OVERLAY? prompt is printed. A negative response causes the plotter to draw the spectrum number axis first. An entry of Y causes the plot without the axis. The program then returns to the COMMAND? prompt. At this point only the O and R as defined above have any meaning. A RETURN causes the program to return to the RANGE? prompt and forget about all previous specifications.

A summary of all commands is as follows:

Character	Function	When Operable
Q	Initialize or close a Q file of spectrum numbers	Anytime
S	Save a spectrum number designated by the cross hair cursors in a Q file	After a Q file is initialized
B	Save a spectrum number designated by the cross hair cursors in a Q file	After an S command
RETURN	Display next screen load on a CRT	Before final screen load is displayed
O	Overlay the next TICP or EICP specified on the CRT	After final screen load is displayed
RETURN	Forget all information about previously specified TICP's or EICP's and return to the RANGE? prompt	After final screen load is displayed
F	Forget all information about specified TICP's or EICP's except the first, and display it on the CRT	After final screen load is displayed

R	Forget all information about previously specified TICP's or EICP's except the first, and overlay it with the next one specified on the CRT	After final screen load is displayed
H	Plot the last specified TICP or EICP on the plotter	After final screen load is displayed

MASS SPECTRA HISTOGRAMS

For mass spectra displayed on a CRT or plotter the dialogue is as follows:

```
SELECT MODE:  CRTPLT
FILE?  CINTAP
SPECTRUM?
```

The response to the FILE? prompt is the usual data filename. There are three possible responses to the SPECTRUM? prompt. Entry of a spectrum number leads to the additional prompts as follows:

```
SPECTRUM?  2
BACKGROUND SPECTRUM?  1
AMPLIFY BY?
```

If no background subtraction is desired, the user should press RETURN. Entry of a background spectrum number leads to the AMPLIFY BY? prompt. An amplification factor of 1–4 may be optionally entered. This causes multiplication of the background spectrum by the factor before it is subtracted. Generally this feature is not useful, and the user should press RETURN. At this point the spectrum is displayed on the CRT, and the cross hair cursors are activated. Interactive examination may begin as described later in this section. To return to the SPECTRUM? prompt, press RETURN.

In place of the entry of a single spectum number as described above, the user may enter the commands QC or QP. Both commands tell the program that the spectrum numbers for the displays are to be retrieved from a Q file generated by BCLRGC. The QC command causes plots on the CRT, and the QP command causes plots on the plotter. Each command leads to the Q-FILE? prompt:

```
SELECT MODE:  CRTPLT
FILE?  CINTAP
SPECTRUM?  QC
Q-FILE?  QCINTA
```

It is extremely important that the user enter a true Q filename generated by BCLRGC. Simply pressing RETURN causes the program to hang, and there is no apparent recovery procedure. Entry of a non-existent filename causes the program to generate spectrum plots from apparently random numbers that have no apparent meaning.

With the QP command a series of spectra defined by the Q file are plotted on the plotter without further operator action. The plotter program keeps a record of the length of each plot. Therefore, if the pen is set initially one inch to the right of a paper fold, each new spectrum will be started one inch to the right of a fold. When the Q file is exhausted, the program returns to the system prompt.

With the QC command the first spectrum defined by the Q-file is displayed on the CRT, and the cross hair cursors are activated. Interactive examination of the spectrum may begin, or the user may press the RETURN to observe a CRT plot of the next spectrum in the Q-file. When the Q-file is exhausted, the program returns to the system prompt.

Regardless of how CRT plots are generated, i.e., by entering individual spectrum numbers or with a Q-file, interactive processing is possible when the cross hair cursors are displayed. Four commands have special meaning, namely B, E, R, and H.

The B command is used to blow-up (expand) a section of a spectrum defined by the cross-hairs. To use this command, first place the cross hairs over the smaller of the two masses that define the section to be expanded, and press B. Then place the cross hairs at the larger mass and press B again. The section defined will be expanded to the maximum extent possible. If the larger mass is defined first, the program blows up.

The E command is used to multiply all abundances by a single integer expansion factor. The masses to which the expansion factor will apply are defined by the position of the cross hairs. To use this command, place the cross hairs over the mass that defines the starting point for abundance expansion, e.g., mass 300, and press E. This causes the EXPAND BY? prompt to be printed. Enter an integer expansion factor, e.g., 10, press return, and the display will be redrawn with all abundances above mass 300 multiplied by 10.

The R command restores the original spectrum to the screen. The B and E commands are cumulative and act on the current screen display, but R always restores the original spectrum. The H command causes the current

CRT display to be plotted on the plotter. If this is the last spectrum of a Q-file, the program returns to the system prompt. Pressing a RETURN will either cause a CRT display of the next spectrum in the Q-file, or a return to the SPECTRUM? prompt if not in the Q mode. Any time a prompt is printed on the screen the user may enter CTRL/L and return to the system prompt.

EXTENDED MEMORY OUTPUT SOFTWARE

The program SPRRGC performs very much like BCLRGC except that it requires 16K of core memory, and the TICP or EICP is generated using only the abundances that maximize at each spectrum number. For example, if the mass 149 abundances at spectrum numbers 100, 101, and 102 are 40, 90, and 40 respectively, the abundance at spectrum number 101 would be retained and the abundances at spectrum numbers 100 and 102 would be discarded. The program uses the overlays RGCOV1 and RGCOV2. A minor difference from BCLRGC is that if the full mass range measured is selected for display, the first mass scanned is ignored. This was intended to compensate for some noise caused by some interfaces that gave false abundance data at the first mass measured. The program SPRRGC will not function correctly if saturated peaks are present in the data file.

The program SPRPLT, and its overlay SPRVL1, display spectra analogous to CRTPLT, but produce only mass maximized spectra as described above. One difference is that the RESOLUTION?: prompt, is given and the only legal responses are 1, 2, or 3. The exact meaning of this resolution is not clear. There is no provision for background subtraction, but the QC and QP commands are enabled as are the interactive options described under CRTPLT.

CHAPTER 5
COMPOUND IDENTIFICATION

The compound identification chapter is not intended to give complete, step-by-step instructions on how to identify organic compounds by mass spectrometry. That would be virtually impossible. The purpose of this chapter is to set into perspective the techniques that are available for compound identification. Major emphasis is on quality assurance which is of utmost importance in the environmental field, and guidelines for accurate identifications are presented to assist the users of GC/MS systems.

5.1 INTERPRETATION FROM THEORY

Interpretations of mass spectra for compound identification may be attempted based on the theory of mass spectra and the rules of fragmentation of organic ions in the gas phase. Unfortunately, this approach is frequently slow and tedious and generally requires considerable experience. It is often difficult to sustain the necessary accurate deductive reasoning process for the long periods of time required to make a large number of correct identifications. Finally, not enough is known about the detailed processes that occur in the fragmentations of organic ions, and this severely limits the approach. However, when all else fails, this approach is clearly justified in important situations, such as an enforcement action or a fish kill.

This method will not be considered further in this manual. However, a number of references to excellent books are included in the bibliography, and many of these discuss interpretation of mass spectra.

5.2 EMPIRICAL SPECTRUM MATCHING

The purely empirical method of searching a file of reference mass spectra to find a similar or exact match of an experimental mass spectrum has been under development for a number of years. Any empirical search and match system has two fundamental components: a data base, which is nothing more than an organized collection of reference spectra, and a system to search the database. Databases may be in printed or machine readable forms and countless search systems are possible. In this chapter several printed spectral collections are recommended and computerized spectral search systems are discussed.

SEARCHING PRINTED DATABASES

Manual searching of printed databases was developed long before computerized systems, and some elaborate indexing schemes were invented to facilitate the user's search. Nevertheless, all manual search systems are rather slow, subject to human error, and intellectually fatiguing. It is also difficult and expensive to update the database since the index usually requires complete revision. In spite of these limitations, this method may be rewarding for small numbers of unknowns if a printed database is available.

For manual searches, the two most useful mass spectral catalogs are the "Eight Peak Index of Mass Spectra", published by the Mass Spectrometry Data Centre (MSDC), Aldermaston, England, and the "Compilation of Mass Spectral Data" (10 peak index) by Cornu and Massot. (See the Bibliography for further information.)

The "Eight Peak Index" consists of three tables in three volumes (31, 101 spectra):

Volume 1.

Table 1. Spectra are arranged in ascending molecular weight order, subordered on number of carbon atoms, hydrogen atoms, etc.

Volume 2.

Table 2. Spectra are arranged in ascending molecular weight order, subordered on m/e values in order of decreasing abundance.

Volume 3.

Table 3. Spectra are arranged in ascending m/e value order, each given where present as first, second, and third most abundant; subordered on other m/e values in order of decreasing abundance.

An enlarged edition of the Compilation of Mass Spectral Data (10 peak index) was published in 1975 in two volumes with three sections. This edition contains mass spectral information on 10,000 compounds.

Part I. Spectra are ordered by molecular weight, giving the compound name, reference number in the original collection, molecular weight, molecular formula, and a listing of the 10 most intense peaks and their abundances.

Part II. Spectra are ordered by molecular formula (in order of C,H,D,Br,Cl,F,I,O,P,S,Si, and others) and includes the same information as Part I.

Part III. Spectra are ordered by frangment ion values in order of increasing m/e values, and includes the 10 most abundant peaks.

The "Eight Peak Index" is usually examined first, because it contains three times as many spectra as the "Compilation of Mass Spectra". In both catalogs, the fragment ion index is the most useful. Detailed instructions for use of these catalogs is beyond the scope of this manual, but these instructions are included in the respective catalogs.

Perhaps the most difficult part of manual searching is the judgment of how well a reference and experimental spectrum match. Most spectra in the catalogs were measured with magnetic deflection mass spectrometers. There is a widespread belief that spectra measured with quadrupole and other types of spectrometers do not agree with spectra from magnetic deflection spectrometers. However, this problem is not at all serious when the spectrometer is tuned according to procedures in chapter 2. With proper ion abundance calibration, quadrupole measured spectra often agree within a few percent with magnetically measured spectra. Other factors, such as differences in inlet systems, should cause greater spectral differences, but these are difficult to recognize in spectral collections.

Comparisons of printed lists of masses and abundances are especially tedious and it is recommended that final decisions concerning matches be made with the histogram plots in the "Registry of Mass Spectral Data", Volumes I-IV, by Stenhagen, et. al. There is also available the "EPA/NIH Mass Spectral Data Base" published by the National Bureau of Standard's Office of Standard Reference Data. This presents the 25,566 spectra of the NIH-EPA collection in bar graph form. The spectra are ordered by molecular weight and each includes the molecular formula, molecular structure, Chemical Abstracts Service registry number, and compound name. The publication is designated NSRDS-NBS 63 and the first supplement will contain about 8000 additional spectra.

REMOTE COMPUTERIZED SEARCH SYSTEMS

The application of computers overcomes some of the problems of manual searching, but computerized search systems are constrained by the size and validity of the database, the thoroughness of the searching algorithms, and the cost of using the systems.

An excellent mass spectral search system (MSSS) is available on a commercial time sharing system at a moderate cost. This MSSS was developed with the joint and cooperative support of the Mass Spectral Data Centre - an agency of the British Government, the National Institutes of Health, the U.S. Environmental Protection Agency, and several other agencies. The principal feature of the MSSS is that it provides different search options and a single database in one convenient system. Table 5.1 lists the many different search options currently available.

For all of the spectral search options in Table 5.1 except KB and PBM, the user enters mass, abundance, or other data from a keyboard/printer or keyboard/cathode ray tube (CRT) terminal that is interfaced to a conventional voice grade telephone line. This terminal may be completely independent of any GC/MS data system. Data entered by the user are transmitted to the time sharing computer that searches the database and prints interim search results at the user's terminal in a few seconds. The user employs these interim results to make judgments about additional data to be entered to improve the search results. This interactive, iterative procedure may continue until the user is satisfied that the database was thoroughly searched. The usual result is that a small number of spectra are found and reported by identification numbers and names. The final choice among these is the responsibility of the user. Complete spectra may be printed at the user's terminal or plotted on a graphics device.

With the KB option, a completely different mode of operation is employed. This is not an interactive, iterative option, but a number crunching system to find the best overall match using the Biemann search algorithm. Most importantly, with the KB option data may be transmitted automatically from the mini-computer of the GC/MS data system to the time sharing computer or entered from a keyboard. The remote computer conducts a search for a match based on the transmitted mass and abundance data, and sends the results back to the mini-computer in a matter of seconds. Direct, computer controlled data transfer requires special communications hardware and several programs for the GC/MS data system. The hardware and programs are available for a number of mini-computers commonly used in GC/MS data systems.

The choice between automatic transmission or one of the keyboard entry options depends on the circumstances. For a few spectra or for special feature searches, e.g., by molecular weight, the keyboard entry option is best. For a large number of spectra, the KB option with automatic data transfer is preferred. The user makes the choice after calling the time sharing system and may change options repeatedly during a time-sharing

Table 5.1. User Options Available with the Mass Spectral Search System

Command	Function
PEAK	Finds all spectra that contain masses and abundances that are designated by the user
MW	Finds all spectra of compounds with a molecular weight that is designated by the user.
MF	Finds all spectra of compounds with a complete molecular formula that is designated by the user.
KB	Computer to computer transmission of spectral data for matching; best for large amounts of data; Biemann algorithm
PMW	Combines peak and molecular weight searches; powerful if you know the molecular weight.
PMF	Combines peak and molecular formula searches; powerful if you know the molecular formula.
MWMF	Combines molecular weight and molecular formula searches; fastest way to find the spectrum of a known compound.
SIM	Computes a similarity index between two sets of data; automatic on KB search.
SPEC	Prints all masses and abundances from spectra in the file; user must supply an identification number.
FICHE	In manual mode, prints fiche numbers and matrix positions of spectra in the file; user must supply an identification number; also in computer mode may drive one model automatic fiche viewer.
PLOT	Plots mass spectra from the file on a graphics device; user must supply an identification number.
COM	User may enter comments, complaints, and suggestions.
EXIT	Causes the MSSS to fade away
AUTH	Searches for literature references by specific author.
COMP	Searches for literature references by specific compound.
ELE	Searches for liteature references by specific elements.
HELPxxxx	Prints explanation of the option xxxx.
INDEX	Searches literature references by specific index terms.
LAB	Calculates the enrichment of isotopically labeled compounds.
META	Calculates the m/e values of various metastable ions that correspond to a given parent/daughter pair.
MOLFOR	Calculates all possible molecular compositions that have approximately the same accurate mass.
PBM	A reverse mass spectral matching program..

Command	Function
PF	Finds all spectra of compounds with a partial molecular formula that is designated by the user.
PA	Searches for proton affinities
SUBJECT	Searches for literature references by specific subject.
OPT,HELP	Prints this list of options

session. Another option which attempts to find the best overall match is PBM, or probability based matching. Presently this requires keyboard entry of the entire spectrum, and its use is limited by this requirement.

If the automatic transmission KB option is selected, the user must run a spectrum abbreviation program on the mini-computer before calling the MSSS. This program, which selects the two most abundant ions in each 14 amu interval, is used as follows:

```
SELECT MODE:  BRVCCP
STORAGE FILE:  SELECT A NAME FOR YOUR
      ABBREVIATED SPECTRUM FILE
SPECTRUM FILE:  THE NAME OF THE FILE CONTAINING
      THE SPECTRA YOU ARE ABBREVIATING
SPECTRUM NUMBER:  46
BACKGROUND SPECTRUM:  43
MINIMUM VALUE:  press return
ANOTHER SPECTRUM:  YES or NO depending on how
      many spectra you are abbreviating
```

if YES —cycle will repeat enabling you to put a number of abbreviated spectra in one file

if NO —the prompt will be ANOTHER FILE? which enables you to abbreviate spectra from another data file stored on the same disk.

If NO again — system returns to SELECT MODE:

After generation of the abbreviated spectrum file another program is used to transmit the data. This program, named MDIREK, uses the Tektronix 4010 or 4012 or 4014 communications interface and will not work with the Digital Equipment Corporation KL8E communications interface. The BAUD rate may be 110–9600 and the program will work with any rate your modem handles and the time sharing system supports. The program is used as follows:

SELECT MODE: MDIREK
SPECTRAL DATA COMMUNICATION PROGRAM
FILE?: NAME OF STORAGE FILE SELECTED ABOVE.

Place the communications interface three position switch in the middle position, select a BAUD rate, full duplex, and call the time sharing system. The standard login procedure is used and a KB is given in answer to OPTION?:

OPTION?: KB
COMPLETE SPECTRUM SEARCH
MAIN FILE(Y OR N)? Y
INPUT TWO TITLE LINES FIRST

Place the switch in bottom position and enter two CTRL/P's followed by a CTRL/A. The first CTRL/P causes the file name, spectrum number, and background spectrum number to print on the CRT. The second CTRL/P causes the sample title to be printed. CTRL/A halts the PDP-8. Place the switch in the top position and press continue. At the end of the spectrum transmission, DATA OK? will be printed. It is important at this point to place the switch in the middle position so that the next spectrum to be transmitted is not disturbed. If the data is acceptable enter a Y; if not enter a N and the data may be edited. Upon completion of the search, the system will transmit all hits with their similarity indexes. It will then prompt with CONTINUE? If the response is NO, you will exit from the KB search. If the response is YES, the prompt will be SAME SPECTRUM (Y or N). Continue as above and transmit the next abbreviated spectrum, i.e., if you abbreviated a second spectrum in the same file you may send it by repeating the sequence of steps beginning with placement of the switch in the bottom position and entry of two CTRL/P's and a CTRL/A.

MINICOMPUTER SEARCH SYSTEMS

An alternative to the large, remote, time sharing system is the use of a search and match system on the mini-computer of a GC/MS data system. Several data system manufacturers offer this kind of system and its use may offer important time and cost savings. The most serious limitations of local database searching are the size of the database, the quality of the database, the difficulty of updating the database, and the very limited searching capability usually offered with these systems.

The EPA minicomputer search system operates as an integral part of the PDP-8 GC/MS system software. The hardware required to use this system is a PDP-8 with 8K of core memory, an extended arithmetic element

(EAE), two disk drives, and a printer such as a teletypewriter, Decwriter, or line printer. An alternative search program operates in 16K and reduces search time by about 5–10 seconds per search. This is recommended for searching very large numbers of spectra, e.g, to match 100 spectra in 8K requires about 17 minutes longer than matching 100 spectra in 16K.

The files required to use this system are not all listed in the disk directory even though one or more of them may be present. In addition, some files are write protected so they are less likely to be clobbered by a user error. Finally, it is required that certain files be on certain disks in specific drive units, or all will fail. The files required and their directory status, write protection status, and disk requirements are listed below.

BMATCH and BMNBVT. These files should be copied on all disks that may be used to store data files. They are conventional directory listed files that are not write protected. The program BMATCH is used for spectrum abbreviation and BMNBVT is an overlay. Spectrum abbreviation is the same function as accomplished with the program BRVCCP when using the KB option and the remote search system. However, its capabilities are significantly enhanced by the IGOOS software described in section 4.3.

BMN8K and BMN16K. These are the actual search programs in 8K and 16K versions. They are conventional directory listed files that are not write protected. They should be on only one disk, plus a backup, which is a special skeleton system disk stripped of the normal data acquisition and data output software. This skeleton system is initially placed in disk unit 1, then moved to disk unit 0 after spectrum abbreviation is finished. The file of abbreviated spectra is also placed on this disk.

NAMES. This file is very special. It is contained on the special skeleton system disk, but it will not appear in the disk directory, and it is write-protected. It occupies blocks 10000–31277. Therefore, these blocks are forever inaccessible and any attempt to write another file into them will result in failure. The number of files on the special skeleton system disk must be kept small and not exceed block 7777. The NAMES file contains Chemical Abstracts nomenclature for all compounds in the database, a unique Chemical Abstracts Registry number for each compound, a molecular formula, molecular weight, and a purely arbitrary identification number.

PRNTSP. This is a conventional directory listed file that is not write-protected. It is on the skeleton system disk and may be used to print the masses and relative abundances of abbreviated spectra as they appear in the database. Please note that the database contains only abbreviated mass spectra.

PRESEARCH. This is a special file that is contained on a disk that has no operating system and cannot be started with the bootstrap loader. This disk is always placed in disk unit 1 after the abbreviated spectrum file is generated on the skeleton system and the skeleton system is moved to disk drive unit 0. This file contains the mass of the base peak for each spectrum and other search algorithm related information. This file is write-protected.

ABBREVIATED SPECTRUM FILE. This is a special file that is on the same disk as the PRESEARCH file, and is write-protected. It contains mass and abundance data for the two most abundant ions in each 14 amu window for 25,555 mass spectra. No two spectra have the same Chemical Abstracts Registry number.

DATA FILES. When using the search system it is strongly recommended that the user have descriptive (up to 64 characters) titles associated with each data file. Abbreviated spectrum files may contain information from one or several data files, and descriptive titles will improve the readability of the printed report form BMN8K or BMN16K.

Procedure (PDP-8).

1. Place the disk containing the data file(s), BMATCH, and BMNBVT in disk unit 0. Place the skeleton operating system in the disk unit 1. Run BMATCH when both disks are up to speed as described below using either Q file or keyboard entry of spectrum numbers. The Q file spectrum number entry mode is strongly recommended for large numbers of spectra and is described first:

```
SELECT MODE:  BMATCH
STORAGE FILE?  TURKEY
FILE?  CINTAP
SPECTRUM?  Q
Q-FILE   QCIN
FILE?  CINTAP
SPECTRUM?  Q
Q-FILE?  Q2CIN
FILE?  6SPK
SPECTRUM?  Q
Q-FILE?  Q26SO
FILE?  press return
```

The storage file is the file that will contain abbreviated spectra from the data files. Each abbreviated spectrum requires three disk blocks. The storage file will be generated on the skeleton system in disk unit 1. The FILE? prompt refers to the data file on disk 0 that contains spectra to be abbreviated. A response of Q to the SPECTRUM? prompt leads to the Q-FILE? prompt. The user must enter the name of a Q file created by BCLRGC using the

IGOOS described in section 4.3. This Q file contains a list of spectrum numbers, and BMATCH will automatically retrieve each, subtract background if called for, abbreviate the spectrum, and place the latter in the storage file on disk unit 1. When the Q file is exhausted, the program returns to the SPECTRUM? prompt. The user may repeat the same process with another data file and the corresponding Q file, or with the same data file and a different Q file that also refers to the data file. Both these options are illustrated in the sample dialogue. Please note that all abbreviated spectra from several data files will be placed in the same storage file on disk unit 1.

After all spectra from all files have been abbreviated, the user must press RETURN in response to the FILE? prompt. If CTRL/L is entered, the program returns to the system prompt, and all abbreviated spectra will be lost.

For keyboard entry of spectrum numbers the user simply enters a spectrum number in place of the Q command as in the following sample dialogue. If no background subtraction is required, press RETURN.

```
SELECT MODE:  BMATCH
STORAGE FILE?  VOLS
FILE?  STD1
SPECTRUM?  56
BACKGROUND?  53
SPECTRUM?  92
BACKGROUND?  88
SPECTRUM?      117
BACKGROUND?  114
SPECTRUM?  127
BACKGROUND  124
SPECTRUM?  192
BACKGROUND?  189
SPECTRUM?  216
BACKGROUND?  212
SPECTRUM?  246
BACKGROUND?  241
SPECTRUM?  press return
FILE?  press return

SELECT MODE:
```

It is important to recognize that Q file and keyboard entry of spectrum numbers may be mixed randomly by the user and all abbreviated spectra will appear in the same storage file on disk unit 1. If spectrum numbers are being entered by the keyboard, the user must press RETURN twice to return to the system prompt as in the example above.

2. Remove both disk cartridges and move the skeleton system from unit 1 to unit 0. Place the disk containing the PRESEARCH and 25,555 spectra in unit 1. Bring both cartridges up to speed.

3. Run BMN8K or BMN16K as shown in the sample dialogue. The FILE? prompt refers to the storage file for the abbreviated spectra on the skeleton system. The user should switch from the CRT to console printer to allow the printed report to appear on the printer.

```
SELECT MODE:  BMN16K
FILE?   TURKEY
```

The printed report consists of the title of the file from which the spectrum was taken, the spectrum number and background subtracted spectrum (if any), the number of presearch hits (base peak only in presearch file), and a list of best hits. For each hit the following is printed:

a. Similarity index (SI) – see discussion of similarity index interpretations in this chapter.

b. The Chemical Abstracts name – this nomenclature system will often generate very long names for commonly known substances, e.g., dieldrin. Future versions of the NAMES file will include common names.

c. The molecular composition – this is printed after the name.

d. The molecular weight – this is printed after the composition.

e. The Chemical Abstracts Registry number – this is a unique number that permanently identifies the compound. This should be included on all reports of the occurrence of this compound to allow users of reports to readily look up the properties of the compound in other printed or computerized data bases.

f. An arbitrary spectrum identification number that has no meaning beyond this search system.

A sample report is as follows:

```
STD #1
SPECTRUM 0056 - 052
  47    HITS

SI=0.589
ETHANE, 1,1-DICHLORO-C2H4CL2    98    75343
#       301
```

```
SI=0.250
ETHANE, 1-BROMO-2-CHLORO-C2H4BRCL    142    107040
#      1121

SI=0.195
CARBONIC DICHLORIDE CCL2O    98    75445
#      307

SI=0.153
ETHENE, (METHYLSULFONYL)-C3H6O2S    106    3680022
#      7360

SI=0.135
ETHANE, 1-CHLORO-2-NITRO- C2H4CLNO2    109    625478
#      3569

SI=0.115
ETHANEDIOYL DICHLORIDE C2CL2O2    126    79378
#      460

SI=0.111
PROPANE, 1,1,2-TRICHLORO- C3H5CL3    146    598776
#      3247
```

When the abbreviated spectrum file is exhausted the program returns to the system prompt. Clearly this search system is oriented to processing large numbers of spectra in an overnight or overlunch batch mode.

One optional program on the skeleton system allows the user to print mass and abundance data from the 25,555 abbreviated spectra file. The NUMBER? prompt refers to the arbitrary spectrum identification number. Pressing RETURN at this prompt causes the program to return to the system prompt.

```
SELECT MODE:  PRNTSP
NUMBER?  301

ETHANE,  1,1-DICHLORO-  C2H4CL2      98      75343
MASS     INT MASS     INT MASS     INT MASS     INT
   14      1   15       1   26       6   27      40
   61      6   63     100   65      31   83      14
  105      1
```

MASS	INT	MASS	INT	MASS	INT
35	1	47	1	60	1
85	9	98	10	100	7

NUMBER?

5.3 QUALITY CONTROL IN COMPOUND IDENTIFICATION

The importance of analytical quality control in organic pollutant analysis cannot be overestimated. Data generated in surveys are being used to set standards for drinking water, air quality, surface water quality, and effluents. Possible correlations between the presence of organic contaminants in drinking water and air and human health effects are under widespread study. In the past many carefully conducted measurements were not documented with sufficient data to support their reliability. This caused doubt about the validity of the measurements and concern for the correctness of correlations and proposed standards.

Several aspects of quality control were emphasized in previous chapters of this manual. Reagent and glassware control is required to minimize the introduction of contamination from the materials used in the sample preparation procedures and this was emphasized in the sample preparation methods in chapter 3.

Instrumentation control is required to assure that the total operating GC/MS system is calibrated and is in proper working order. If a computerized GC/MS system is used to collect data, the computer data system must be included in the performance evaluation. The recommended instrumentation control procedure was described in section 2.6. This employs a standard reference compound and a set of reference criteria to evaluate the performance of the overall system. This evaluation should be performed on each day the GC/MS system is used to acquire data from samples or reagent blanks. The records from the performance evaluations should be maintained with the sample and reagent blank records as permanent documentation supporting the validity of the data. The major emphasis of this section is the quality control that applies to compound identification.

EVALUATION OF BLANKS AND REAGENT BLANKS

The blank is an experiment that is required for all samples. A reagent blank is defined as an experiment that employs all procedures, quantities of materials, glassware, etc. used in the sample preparation except that no water, air, tissue, or sediment sample is used. A low organic water blank is recommended in place of a reagent blank in several procedures, e.g., inert gas purging and trapping. One of these blanks is required even when contamination from glassware and reagents is well controlled. The data from the blank is the documentation that proves that good control was exercised, and it defines the level of background that was beyond control. The evaluation may be a straight forward comparison of corresponding peaks and mass spectra in the blank and sample.

An effective but rapid technique for comparison of blanks and samples employs the extracted ion current profile (EICP) of one or several ions. An EICP is defined as a plot of the change in relative abundance of one or several ions as a function of time. The data for this plot is extracted from all the ion abundance measurements made over the mass range observed during the elution of the separated components from the GC. The EICP produces an apparent increase in sensitivity by subtracting from the total ion current profile all the ion abundance data that is contributed from background, unresolved components, and other irrelevant ions. The EICP generator is a standard data reduction program on all modern computerized GC/MS systems and is described in chapter 4 and section 6.1. A fast graphics display device (CRT) is required to facilitate reviewing a large number of EICP plots.

It is emphasized that it is not necessary to have even a tentative identification of a compound to apply this technique to blank evaluation. To conduct an EICP comparison, mass spectra of all peaks in the sample are examined. One or several ions that are prominent in a spectrum from each peak are selected, and sample and reagent blank EICP's are generated on the CRT. In most cases comparison of these EICP's permits straightforward judgments concerning the presence of compounds in the sample and the blank. Figure 5.1 shows EICP's for mass 171 from a sample and from the corresponding reagent blank. Clearly a compound having a mass 171 ion is present in the sample, but no corresponding peak is observed above the noise level in the reagent blank. Figure 5.2 shows a sample and reagent blank that contain the same compounds by EICP analysis. For a valid comparison this technique does require exactly the same GC and data collection conditions for the sample and blank. It is possible that the compounds in the blank and sample that elute at the same time are different, but contain the same ion or ions. If this is suspected, the entire spectrum from each must be examined to confirm or deny this possibility.

If the concentration of a blank component equals or exceeds the concentration of sample component, the decision is clear and the compound must not be reported. A far more difficult judgment must be made when the concentration of a sample component exceeds its concentration in the blank. The material could, of course, be a true sample component.

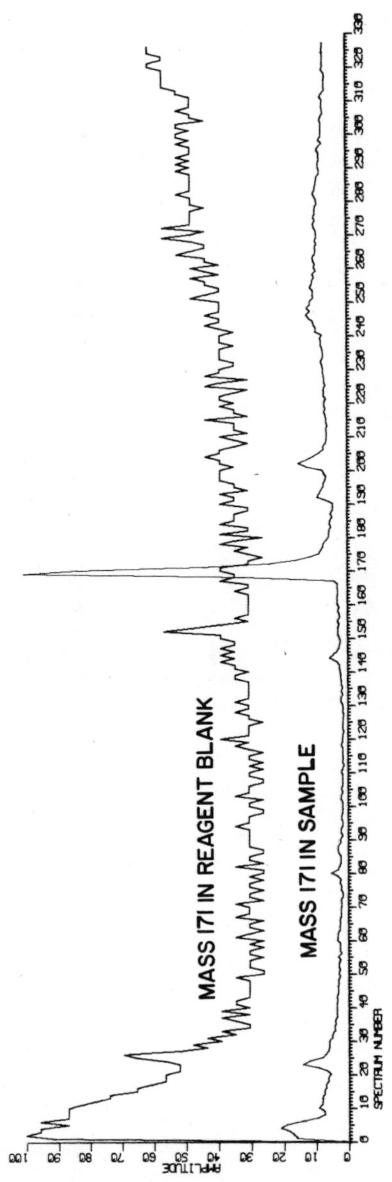

Figure 5.1 The extracted ion current profiles for mass 171 from a sample and the corresponding reagent blank.

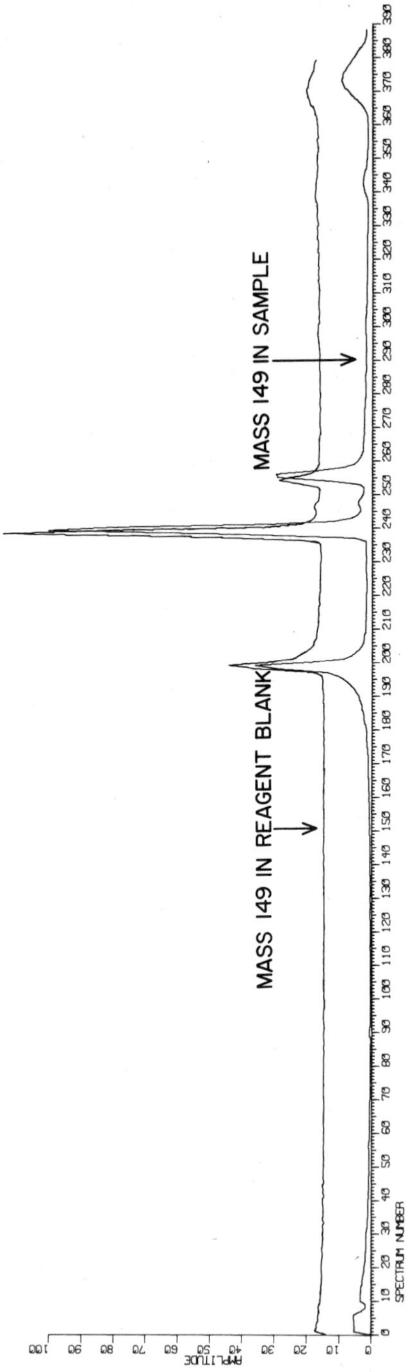

Figure 5.2 The extracted ion current profiles for mass 149 from a sample and the corresponding reagent blank.

Alternatively it has been observed empirically that compounds in the blank sometimes merely appear to be at lower concentrations than the same compounds in the corresponding sample. Figure 5.3 shows the total ion current profiles from a sample and a corresponding reagent blank. Careful comparison of the profiles reveals a very similar pattern of peaks and valleys in certain areas, e.g., spectrum numbers 170-190 and 235-245, yet a significantly lower apparent concentration in the reagent blank. There are several possible reasons for this. One rationalization used, in the case of solvent extracts, is that impurities in the solvent of the reagent blank are adsorbed more efficiently onto the drying agent and other surfaces. With extracts containing some water, the wetting effect of the water precludes efficient adsorption on surfaces, and impurities are carried on in the solvent. Alternatively, certain solvent impurities may be lost more readily from the blank than from the sample extract during concentration. Perhaps the general organic background matrix in the sample extract retains the solvent impurities. Both explanations are reasonable but unproved. In view of the uncertainties, any compound that is observed in the sample should not be reported if it is part of an overall pattern of peaks that is repeated in the blank, although at a lower apparent concentration. This same overall pattern will usually persist in the acid, neutral, and basic fractions of a solvent extract.

SUPPORTING EXPERIMENTS

Chemical ionization, field ionization, and high accuracy mass measurements are GC/MS techniques that are capable of generating very strong evidence in support of identifications. Some of these are discussed in Chapter 6. However, the production of this evidence is restricted because only a relatively few laboratories have developed capabilities with these techniques. High accuracy mass measurements are further limited by sample size, since some sacrifice in sensitivity is required to achieve the high accuracy.

After a tentative identification is made, several other types of supporting experiments become possible. Retention time data from the GC/MS of a pure compound (standard) may be compared with analogous data from the sample component. Similarly, the mass spectrum of the standard, obtained under the same conditions that were used for the sample, may be compared with the sample component spectrum. The standard may be dissolved in water at an appropriate concentration, isolated, and measured. The recovery of this spike in the same fraction in which the suspected component appeared, and the observation of equivalent mass spectra for the spike and the sample component, is strong evidence for confirmation of the identification.

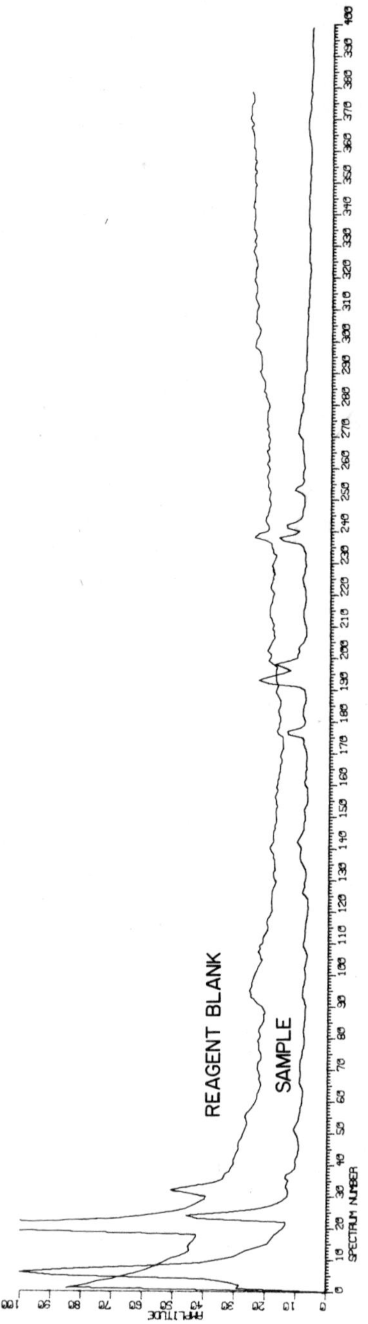

Figure 5.3 The total ion current profiles from a sample and the corresponding reagent blank.

REQUIREMENTS FOR PURE STANDARDS

Clearly the most convincing evidence for an identification is obtained by the examination of pure standards corresponding to suspected sample components. However, the existence of this evidence is constrained by the availability of the pure standard and the additional cost and time required to examine it. Because it is not usually possible to predict which compounds will be found, some standards will not be available immediately. There are many very practical limitations imposed on the development and maintenance of a large library of pure authentic standards. Many compounds are obtainable from fine chemical supply houses, but procurement time is variable and may extend to weeks or months. Some compounds are not available from any supplier, frequently because they are by-products of industrial processes rather than manufactured products. This same limitation of standard availability also precludes calibrated concentration measurements in many cases.

SIMILARITY INDEX INTERPRETATIONS

Because of the problem of standard availability, it is worthwhile to determine whether reliable identifications can exist without standards. One criterion for a reliable identification that might be used is a quantitative measure of the exactness of match between an experimental mass spectrum and a spectrum from the printed literature or a computer readable database. A similarity index (SI), calculated on a scale from zero to one, has been described and used in several computerized mass spectrum matching systems including the KB option and PDP-8 search systems described in this chapter. Experience with this SI indicates that, in general, a value greater than about four tenths corresponds to a reasonable match between two mass spectra. A reasonable match often does imply an identification, but sometimes it does not. Position isomers and members of homologous series of compounds often give very similar mass spectra. Another undetermined number of compounds simply are not uniquely characterized by their mass spectra. Figure 5.4 shows the mass spectrum of an unidentified compound and the spectrum of the compound chloropicrin, Cl_3CNO_2. The match is clearly good by inspection and a SI of 0.453 was calculated. Nevertheless, the unknown whose spectrum is in Figure 5.4 is not chloropicrin as determined by the gas chromatographic behavior of the unknown and pure chloropicrin.

Another problem with identifications based on empirical spectrum matching is that significant differences in ion abundance measurements are sometimes observed when the mass spectrum of a compound is measured on two different spectrometers. Some of these differences are probably caused by non-uniform calibration procedures or by a failure to use an ion

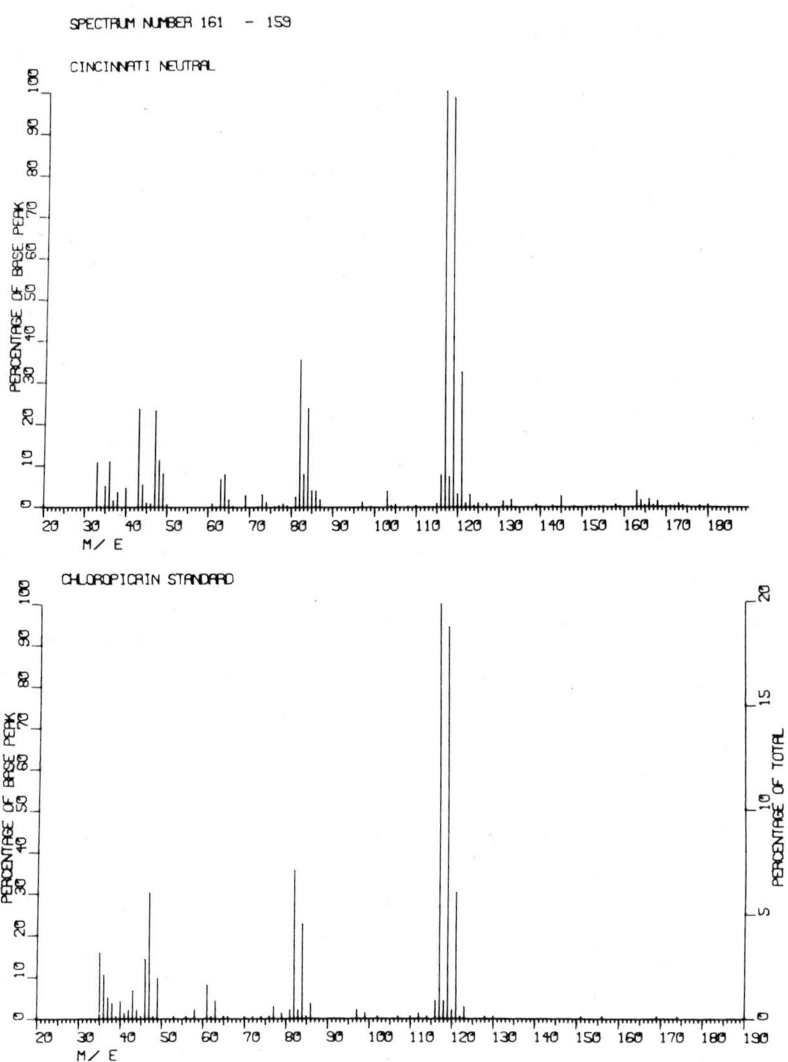

Figure 5.4 The mass spectrum of an unidentified compound and the mass spectrum of chloropicrin.

abundance calibration procedure. Different types of inlet systems also may have significant effects on relative abundance measurements. A GC or batch inlet system that is operated in the 100-250C temperature range may promote temperature dependent fragmentations and reduce abundances of molecular and other high mass ions. With a well designed direct inlet system, these temperature effects may be largely precluded. As a result of these factors, it is quite common for two spectra of the same compound, that were measured with different inlet systems or spectrometers, to give a rather low SI. A low SI may also be caused by unresolved or partly resolved components which generate mass spectra containing extraneous ion abundance measurements.

It is concluded that the SI must be used with caution. A relatively high SI may be regarded as an indication of a reasonable match, but only as suggestive of the probability of an identification. A relatively low SI cannot be regarded as complete rejection of a possible identification.

THE QUALITY INDEX

Another criterion for reliable identifications when standards are not available is based on an assessment of the quality of the ions in the experimental and reference mass spectra. In the SI calculation, molecular ions (M^+), molecular ions having naturally occuring isotopes (typically M + 1, Cl, S, etc.) and all key fragment ions are weighted the same as many very common fragment ions. However, the M^+ ion, for example, is unique in every mass spectrum and has significance to an identification that far outweighs most other ions. Mass spectra may be categorized according to the quality of the ions observed and a quality index (QI) calculated that is a weighted similarity index.

$$QI = SI * W$$

Several categories of quality are as follows:

(a) If the molecular ion is observed and the observed distribution of abundances for it and its isotope containing species are within 10% of the expected distribution, the W is 1.0. The significance of the SI value is considerably enhanced.

(b) If a molecular ion is observed but the isotope data are not within 10% of the expected value, lower confidence is assigned by a W of 0.75.

(c) Failure to observe a molecular ion but the observation of key fragments that account for all the atoms of the molecular ion suggests a W of 0.5. This factor may be raised or lowered between 0.4-0.6 depending on the observation of consistent isotope data in the key fragment ions.

(d) The lowest confidence is placed on spectra which do not contain adequate fragment ions to account for all the atoms of the molecular ion. A W of 0.1 is assigned to these spectra.

It is recognized that position isomers may not be distinguishable under any circumstances, but this is often true even when pure standards are available.

The quality index is amenable to additional positive adjustments by 0.1-0.2 QI units if all major fragment ions are scrutinized and found to be reasonable and compatible with the assigned structure. Reasonableness should be based on compatibility with the accepted principles of fragmentation of organic ions in the gas phase. With magnetic deflection spectrometers additional fractional quality points may be added if fragmentations are supported by the observation of ions from the decomposition of metastable ions.

Good spectrum matches that have a QI of 0.75-1.0 are considered reliable identifications when pure standards are not available.

CHAPTER 6
ADVANCED ANALYTICAL TECHNIQUES

The purpose of this chapter is to describe several GC/MS techniques that are of a more specialized nature than the broad spectrum methods described in the first five chapters. Each of these specialized techniques has achieved prominence in mass spectrometry, and each is expected to become more important in future applications of GC/MS.

Selected ion monitoring is a real time technique whose operational philosophy differs sharply from the broad spectrum approach emphasized in the previous chapters. Unlike the broad spectrum approach, selected ion monitoring requires knowledge of the compounds that are to be measured. In a very real sense it is the application of a mass spectrometer as a super selective and sensitive GC detector. Its selectivity far exceeds any known GC detector, and its sensitivity is at least the equal of the most sensitive detectors known. The technique is discussed in section 6.1.

During the last fifteen years the emphasis placed on qualitative analysis by mass spectrometry has somewhat obscured the quantitative analytical strengths of the technique. Today quantitative analysis by mass spectrometry is more significant than ever, especially when used with selected ion monitoring. Quantitative analytical methods are discussed in section 6.2.

The development of open tubular column GC technology is expected to be very significant during the next five years, and reliable, long lasting precoated columns will likely be available at relatively low cost from a variety of suppliers. This technique is changing so rapidly that section 6.3 is devoted largely to a presentation of a few basic principles, and to a discussion of several potential areas of application.

Chemical ionization GC/MS is discussed in section 6.4, and because of the very broad scope of this relatively new technique, the presentation is limited to a few basic principles, quality control considerations, and environmental applications.

Precise mass measurements are currently beyond the capabilities of most environmental GC/MS laboratories. The brief discussion of this technique is included in section 6.5 primarily for the information of readers who should be aware of the capabilities that exist, and the advantages and disadvantages of the method.

6.1 SELECTED ION MONITORING (SIM)

The computer controlled mass spectrometer has two general modes of operation as a continuous detector in chromatographic systems. One mode is to acquire conventional mass spectra of components as they emerge from the chromatographic system. These mass spectra are used to identify the individual components. This broad spectrum approach was emphasized in the first five chapters of this manual.

The alternative mode is to apply the mass spectrometer as a substance selective detector. This mode is called selected ion monitoring (SIM) which is defined as the dedication of a mass spectrometer to the acquisition of ion abundance data at only selected masses in real time as components emerge from the chromatographic system.

Selected ion monitoring is not a new principle in mass spectrometry. A technique called peak stepping or peak switching has been used for decades in the precise measurement of isotope ratios. In this classical application, the ionic abundances measured were usually limited to those separated by just a few atomic mass units, for example $^{16}0$ and $^{18}0$. In recent years there has been a very significant increase in the applications of SIM. This was brought about largely by the development of the computer controlled quadrupole mass spectrometer as vapor phase chromatography detector. The presence of the gas chromatography inlet system permitted the introduction into the mass spectometer of very small samples of complex mixtures in an easily handled liquid form. The computer controlled quadrupole mass spectrometer provided a method for high speed and very accurate and precise ion monitoring over a very wide mass range. Finally the dedicated minicomputer and its related peripherals gave the experimentalist access to a wide range of control functions and real time ion monitoring techniques.

DEFINITIONS OF TERMS

There is usually a period of confusion in terminology, concepts, etc. any time that technology is rapidly advanced by a number of individuals and organizations in a relatively short time period. The application of a mass spectrometer as a substance selective detector in chromatography is no exception. The terms accelerating voltage alternation, mass fragmentography, single ion detection, and multiple single ion detection are among a number of terms that have been used to describe this technique. An analysis of this terminology led to the recommendation of a standard term, selected ion monitoring (SIM), because it best conveys to the reader the significant information about the technique that sets it apart from other techniques. The term SIM is general in that it does not imply a particular type of spectrometer, the number of ions measured, or the type of ions measured. It must be recognized however that SIM is a real time measurement technique and that a designation is also required for the output obtained from SIM. The name selected ion current profile (SICP) has been recommended as the most meaningful and appropriate. Consistent with the definition of SIM, a SICP is then a plot of the change in ion abundance as a function of time using abundances measured by SIM.

Clear precise terminology is particularly important in computerized GC/MS work because there are several other widely used techniques that may be confused with SIM and the SICP. It is important to understand these techniques to appreciate the overall advantages and disadvantages of SIM. Perhaps the most widely used real time data acquisition technique in GC/MS is the continuous repetitive measurement of spectra (CRMS). Figure 6.1 contains a schematic diagram of CRMS and two types of data reduction that are in common use. The sawtooth diagram in the top of Figure 6.1 is a representation of CRMS during the elution of components from a gas chromatograph. Each solid line represents a sweep of the mass spectrometer from an arbitrary starting mass, e.g., 40 amu, to an arbitrary ending mass, e.g., 400 amu. Each dotted line represents the resetting of the mass spectrometer sweep control to the starting mass. In a typical GC/MS run, several hundred to over a thousand mass spectra may be acquired in this way. Each sweep of the mass range usually requires a time in the range of 2–5 seconds, but faster or slower scans may be used in some cases. The second diagram from the top in Figure 6.1 merely shows that each solid line of the sawtooth represents a mass spectrum as displayed in the standard histogram format. The most important idea is that CRMS produces a set of mass spectra that are more or less complete depending on the selection of the mass range. Each integer mass between the starting and ending masses is measured and recorded. This is in sharp contrast to SIM where measurements are made at only a few masses in real time.

The third diagram from the top in Figure 6.1 illustrates a widely used data reduction process that uses data acquired by CRMS. Each point on the ordinate is the normalized sum of all the ion abundance data in a single mass spectrum and, each point on the abscissa represents the spectrum number or a corresponding unit of time. This plot is referred to as a total ion current profile (TICP) which is defined as a normalized plot of the sum of the ion abundance measurements in each member of a series of mass spectra as a function of the serially indexed spectrum number. This same plot is often referred to as a reconstructed gas chromatogram (RGC), but this nomenclature is not preferred as it does not accurately define a TICP. An RGC could just as well be the output of a flame ionization detector, as redrawn by a draftsman. An important point to recognize is that the TICP in Figure 6.1 is the result of a computer data reduction and not real time display. With magnetic deflection spectrometers it is common to continuously monitor the *unresolved* ion beam and produce a total ion current plot in real time. This plot should be similar to the TICP in Figure 6.1, but clearly it will differ in that it will contain contributions from ions below mass 40 and above mass 400.

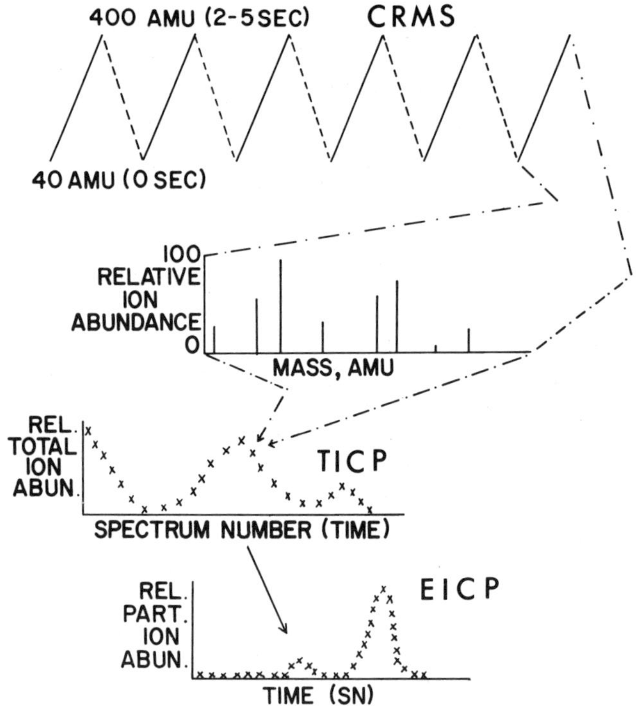

Figure 6.1 A schematic diagram of continuous repetitive measurement of spectra, a TICP, and an EICP.

There is one more very valuable data reduction technique whose output is most often confused with a SICP. In this technique data acquired by CRMS, and perhaps displayed in a TICP, is further reduced by plotting the change in relative abundance of one or several ions as a function of time. This plot is illustrated at the bottom of Figure 6.1. The plot appears very similar to a SICP but the data used are quite different. The significance of this difference is explained below, but first a clear precise name for this output is required. The original name suggested was mass chromatogram. Unfortunately this is not a very descriptive name and it may be confused with the output from a gas density gas chromatographic detector. The name extracted ion current profile (EICP) is more meaningful because the data for the few ions used in the plot are extracted from the larger set used to generate a TICP. The terms limited mass output or limited mass search are often used to describe this same plot. However they are less meaningful than EICP since the nature of a limited mass is not clear.

The significant difference between a SICP and a EICP is that SIM produces a real increase in signal/noise by time averaging random noise. The EICP produces an apparent increase in sensitivity by removing from the TICP the ion abundance data from background, unresolved components, and other irrelevant ions. The contrast between continuous repetitive measurement of spectra and SIM is illustrated in Figure 6.2. If the sweep of the complete spectrum is made in 3.6 sec (3600 millisec) then the data system may integrate the ion current at each integer mass for 10 msec (3600 msec/360 amu). If the same total time is allowed for all the measurements at each of the selected masses 99, 157, 203, and 250 amu in real time, then integration of signal intensity at each may proceed for 900 msec (3600 msec/4 amu). The longer integration time during SIM permits enhancement of the signal to noise ratio by averaging of random noise. Therefore there is a substantial improvement in the detection limit by SIM. This is in contrast to the EICP which still uses the 10 msec data with its inherently lower signal/noise.

The SICP illustrated in the third diagram from the top of Figure 6.2 was generated by summing the abundances of all four ions measured during SIM. Clearly one could also plot the change in abundance of each ion separately and we make no distinction between various types of SICP plots. However, as illustrated in the bottom of Figure 6.2, a SICP for mass 125 would yield no peak since mass 125 was not measured during SIM.

Figure 6.3 is a display of a TICP, SICP, and an EICP. The TICP was generated from CRMS over the 40–400 amu mass range during chromatography of seven chlorobiphenyl isomers. Five nanograms of each isomer was injected and an 11 msec integration time was applied at each mass. The total time for a 40–400 amu sweep was about 5 sec. The EICP

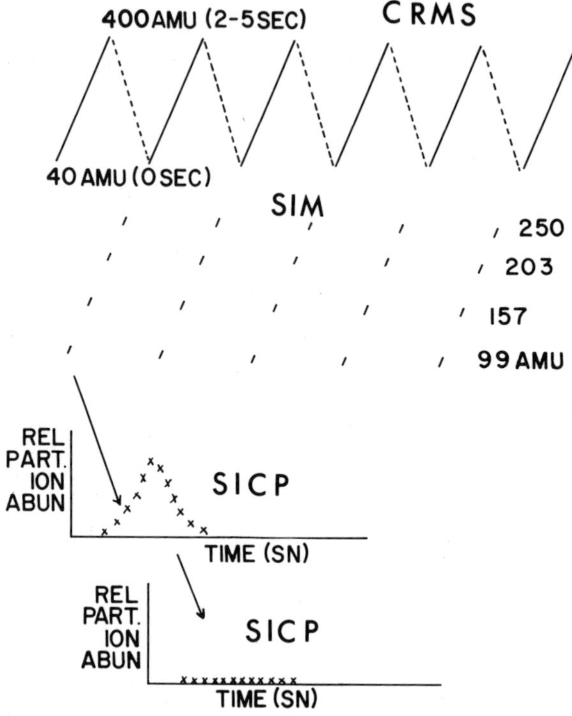

Figure 6.2 A schematic diagram of continuous repetitive measurement of spectra, SIM, and two SICP's

was obtained from the TICP using seven masses characteristic of chlorobiphenyls. The SICP is the result of SIM using the same seven masses, but an integration time of 540 msec on each. The SICP is the sum of the data from the seven masses. The signal to noise contrast between the SICP and EICP is clearly demonstrated. This illustration is not meant to imply that EICP is not valuable technique, but to show the differences in the methodology.

A routine application of an EICP is shown in Figure 4.3. In this example the TICP data has an adequate signal/noise and the EICP was used to effectively to highlight those areas of the chromatogram having abundant mass 149 measurements.

SELECTION OF MASSES

The principle potential advantages of selected ion monitoring may be summarized as follows:

1. High Selectivity
2. Tunable Selectivity
3. Qualitative Reliability
4. High Sensitivity
5. Reduced Need for Sample Purification
6. Quantitative Accuracy

In practice it may not be possible to achieve all of the advantages simultaneously. There is a general tendency to select the most abundant ion or ions in a spectrum in order to optimize sensitivity. In certain cases this may have a significant effect on the selectivity, and therefore the reliability of the measurement. For example, the most abundant ion in the electron ionization spectrum of 2-methylbenzothiazole is the molecular ion, mass 149. However selection of this ion for SIM could result in measurement errors due to the ubiquitous phthalate esters which display intense mass 149 fragment ions. The selection of a less abundant but more selective ion may provide adequate detection limits and preserve the reliability of the measurement.

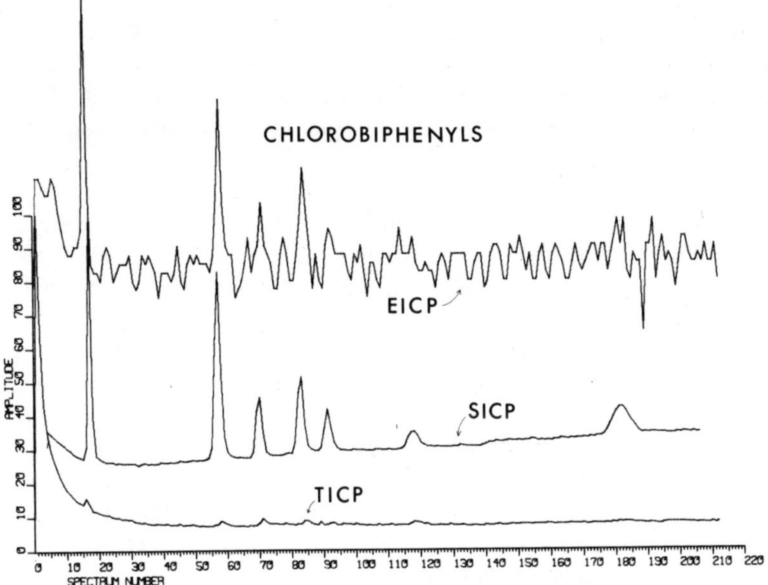

Figure 6.3 A TICP, SICP, and EICP from the chromatography of seven chlorobiphenyl isomers.

One method of selecting masses is to employ a computerized mass spectral search system (chapter 5) to evaluate the selectivity of candidate masses in the spectrum of the compound of interest. In this method, the frequency of occurrence of several candidate masses in a large data base of mass spectra is determined, and the results are used to select a mass or masses having an acceptable trade-off between selectivity and sensitivity.

A method of maintaining qualitative reliability is to monitor several ions from the same compound that have an established abundance relationship. The molecular ion and its corresponding isotope containing species have abundance relationships that are well known. If the compound of interest contains chlorine, bromine, or other elements with several abundant isotopes, an excellent approach to qualitative reliability is to monitor several ions and compare the observed and expected abundance ratios. Figure 6.4 shows a number of calculated chlorine/bromine isotope distribution patterns normalized to the most abundant ion of the group. Within these patterns are numerous possibilities for comparisons of ratios. For compounds that do not contain readily measurable isotopic species, it is recommended that several ions of experimentally determined relative abundance be monitored and their abundance ratios compared.

If several ions from the same compound are monitored, selectivity may be improved significantly by a data reduction program that plots a sum of the abundances of the selected ions at a spectrum number if and only if all the selected ions are above a defined threshold. The advantages of this technique will be more generally recognized as programs with this capability are developed.

The reduction in sample preparation as a result of SIM will depend on the nature of the sample. In the environmental field, air and relatively clean water samples offer the best possibility for elimination of all extract purification. For fatty tissue, sediment, and sewage samples some reduction in extract purification is usually possible.

One difficulty with SIM is the simultaneous measurement of two or more components with significantly different concentrations. Selection of a long integration time will enhance the signal/noise of less abundant ions, but abundant ions will saturate the detection system. Alternatively, a short integration time may avoid saturation of the detector but preclude clear observation of the less abundant ions. A partial solution to this problem is a method for the dynamic selection of integration time as a function of signal strength (IFSS), as described in chapter 2.

SIM USING THE CONTROL PROGRAM

In section 2.7, the control program options were described and the possibility of using this program for selected ion monitoring was implied.

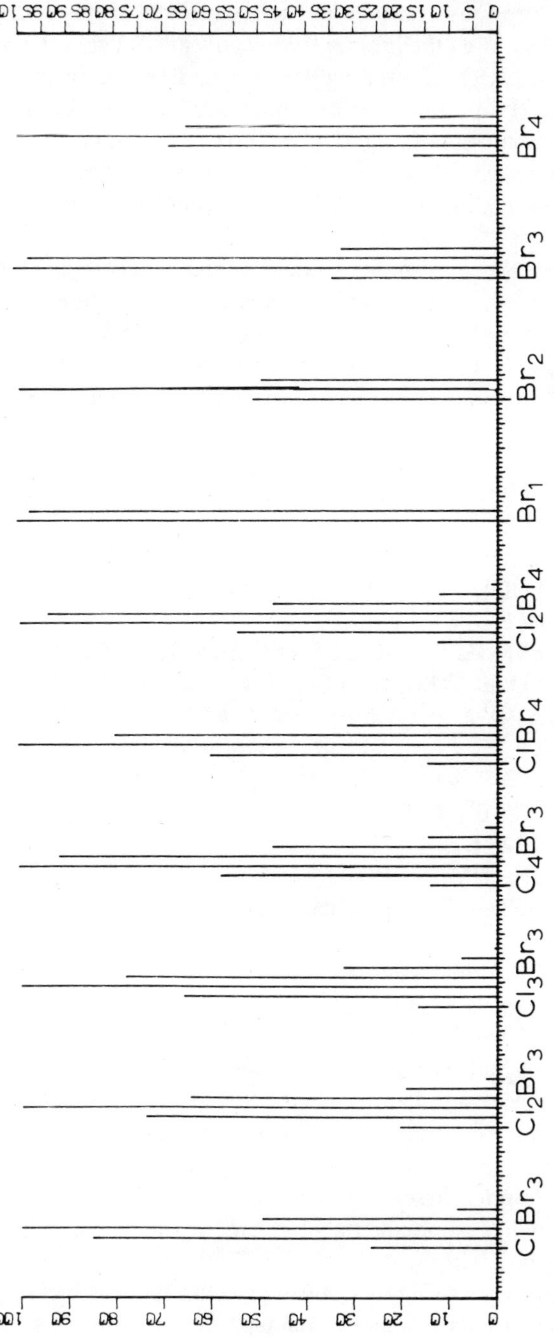

Figure 6.4 Some calculated chlorine-bromine isotope distribution patterns.

However all previous examples of the application of this program in this manual demonstrated the program with conventional wide mass range data acquisition. In place of entering a single mass range in response to the MASS RANGE? prompt, the user may enter up to 8 individual masses separated by semicolons or a mixture of up to 8 individual masses and mass ranges separated by semicolons. The masses and/or mass ranges must be entered in ascending, non-overlapping order and they need not be contiguous.

If more than one mass or mass range is entered, a corresponding number of integration times and samples/amu must be entered separated by semicolons. These have the same meaning as described in section 2.7. The following sample dialogue illustrates selected ion monitoring with the control mode using eight ions characteristic of chlorinated biphenyl compounds:

```
SELECT MODE:  CONT
CALIBRATE?:  N
TITLE:  SIM FOR PCB'S USING 8 IONS
CALIBRATION FILE NAME:  CAL
FILE NAME:  SIM
MASS RANGE:  190;224;260;294;330;362;394;426
INTEGRATION TIME:  400;400;400;400;400;400;400;400
SAMPLES/AMU:  1;1;1;1;1;1;1;1
THRESHOLD:
RT ON CRT?:
RT GC ATTEN:  5
MS RANGE SETTING?:  H
MAX RUN TIME:  30
DELAY BETWEEN SCANS (SECS.)?
```

The ions used in this example are selective for chlorobiphenyl compounds with different levels of chlorination and do not give optimum sensitivity. The choice of 400 milliseconds as an integration time is an example of the use of SIM to signal average random noise for a significant improvement in the signal to noise ratio. It also embodies the idea of choosing an integration time such that a complete cycle through the selected masses requires about the same overall time as a conventional data acquisition from 33–450 amu using CRMS. Under these conditions the spectrum number index displayed on the SICP will be approximately the same as the spectrum number index displayed on the TICP.

The advantages of using the control program for SIM include the ability to output the results using the standard output programs described in section 4.1 and MSSOUT described in section 4.2. Output as a TICP will produce a SICP which displays the sum of the abundances of the selected ions as a function of time. Individual masses may be displayed using the EICP capability. Another significant advantage is the IFSS capability as defined in section 2.7. Sample dialogue for SIM with IFSS is as follows:

```
SELECT MODE:  IFSS
CALIBRATE?:  N
TITLE:  SIM FOR PCB'S USING 8 IONS
CALIBRATION FILE NAME:  CAL
FILE NAME:  SIM
MASS RANGE:  190;224;260;294;330;362;394;426
SAMPLES/AMU:  1;1;1;1;1;1;1;1
MAX RPT COUNT:  16
BASE INTEGRATION TIME:  1
RPT COUNT BEFORE CHECKING LOWER THRESHOLD:  4
LOWER THRESHOLD:  2
UPPER THRESHOLD:  2
RT ON CRT?:
RT GC ATTEN:  5
MS RANGE SETTING?:  H
MAX RUN TIME:  30
DELAY BETWEEN SCANS  (SECS.)?
```

For each mass or mass range selected a sample/amu value must be entered. A single entry is acceptable for the base integration time and the other IFSS parameters. However one may enter a string of values separated by semicolons for IFSS parameters and these will be applied to the masses or mass ranges in the usual way.

SPECIALIZED SIM PROGRAMS

There are several data acquisition and data output programs that are specifically for selected ion monitoring, and cannot be used for more generalized types of data acquisition or data output. However these programs have several advantages that may favor their use in certain applications. The data acquisition programs have a digital integration feature which may be used to improve signal/noise with very low level signals. The data display programs have a digital smoothing feature to further improve signal/noise, and the plotter display program allows the user to exactly overlay selected ion current profiles of two or more ions.

The data acquisition programs are named CRTSIM and PLTSIM. The programs are the same except that CRTSIM creates the real time display on the CRT screen, and PLTSIM creates the real time display on the plotter. The dialogue for the two programs is the same:

```
SELECT MODE:  CRTSIM OR PLTSIM
CALIBRATION FILE:  CAL7
MASS(ES):  198,442,443
INTEGRA. TIME:  50
NO. POINTS:  10
RUN TIME:  30
SCALE FACTOR:  10
TITLE:  SIM TEST
DATA FILE NAME:  DFTPP
DATA
```

In response to the MASS(ES): prompt the user may enter from one to eight masses separated by commas. It is possible to enter fractional masses such as 440.3. An integration time from 1 to 4096 msec may be specified for each mass entered, or a single time may be entered which applies to all the masses. If the number of integration times entered is less than the number of masses, the last time entered will be used for the masses which had no integration time specified. The number of points is the number of times each mass is integrated before a combined value is stored in the data file. The run time is entered in minutes, and the scale factor is an integer amplification factor for the real time data display. During data acquisition three commands are acceptable with the following results:

CTRL/L terminates data acquisition, saves the data file, and returns to the monitor

E terminates data acquisition, deletes the data file, and returns to the monitor

Fn (where n = an integer) changes the scale factor in real time, but does not affect the data already stored in the data file.

The data files created by the two data acquisition programs are compatible with the two data output programs named CRSMPT and PTSMPT. The data files are not compatible with the output programs described in chapter 4 (however see CONVRT below). Furthermore the CRSMPT and PTSMPT programs cannot output data generated by the control program in either the control or IFSS modes. Sample dialogue for CRSMPT is as follows;

```
SELECT MODE:  CRSMPT
FILE?  PFTBA
FIRST MASS?  219
SECOND MASS?  502
INDIVIDUAL NORMALIZATION?  Y
```

This program plots digitally smoothed selected ion current profiles for two ions on the CRT. Only the first 200 data points are displayed for each ion. By pressing any key, the next 200 data points are displayed and this procedure may be continued until all data points have been plotted. The program then returns to the FIRST MASS? prompt. A Y response to the INDIVIDUAL NORMALIZATION? prompt will cause each profile to be normalized on the largest peak in each individual profile. Any response except Y will cause all the profiles to be normalized on the largest peak in the entire data file.

Digital smoothing causes actual plotting to begin at the seventh data point and terminate six data points before the last point in the file. The smoothing may result in the most intense peak for any ion being a few percent too large or small. A CTRL/L command returns the user to the system monitor.

The dialogue for the plotter program PTSMPT is as follows:

```
SELECT MODE:  PTSMPT
FILE?  PFTBA
MASS?  219
SMOOTH?  Y
INDIVIDUAL NORMALIZATION:  Y
OVERLAY?  Y
MASS?  502
SMOOTH?  Y
INDIVIDUAL NORMALIZATION?  Y
OVERLAY?  N
MASS?  503
SMOOTH?  Y
INDIVIDUAL NORMALIZATION?  N
OVERLAY?
```

This program is somewhat different in that only one profile is plotted and then the user has an option of either overlaying one or more profiles (perhaps using a different color ink) or plotting the additional profiles separately. The actual plotting begins immediately after the INDIVIDUAL NORMALIZATION? prompt which has the same meaning as in

CRSMPT. Also, the user has the option of omitting the digital smoothing routine. Again the CTRL/L command returns the user to the monitor.

There is a program that converts CRTSIM or PLTSIM data files into files compatible with the output programs described in chapter 4. The program is called CONVRT and is run as follows:

SELECT MODE: CONVRT
FROM UNIT: 0
INPUT FILE: TEST
TO UNIT: 0
OUTPUT FILE: CTEST

SIMEXC (12K and EAE required) and CRTSM2 (an overlay) expand the capabilities of CRTSIM by allowing an unlimited number of mass sets to be monitored one at a time, with each set containing up to 8 masses. The mass sets entered, together with integration times are withdrawn sequentially, so that users should enter these on the basis of retention time. The user only calls SIMEXC, never CRTSM2, and SIMEXC controls all data flow and program control. After calling SIMEXC, the dialogue is exactly like CRTSIM, except after all information is input for a set of masses the question MORE INPUT? is asked, to which the user should respond Y if additional sets of masses are to be entered. Any other response is treated as a negative one, resulting in the printing of DATA on the screen. One pitfall must be pointed out so that it can be avoided. A different file name is required for data storage for each set of masses. SIMEXC tests for duplicate file names among catalogued files, but does not test for duplicate entries within its own dialogue. This can result in a halt in the program overlay CRTSM2 if duplicate names are entered in SIMEXC. SIMEXC-CRTSM2 detects the presence of a saturated peak during data acquisition. At the time a file is catalogued, it prints SATURATED PEAK on the screen and halts, forcing action by the operator. To continue the run, the CONTINUE switch on the computer must be depressed. Mass sets may be changed in two ways. First, after the run time for a given set has expired, the next set is automatically withdrawn and data acquisition begins for that following cataloging of the previous file. Secondly, striking CTRL-L during data acquisition automatically halts acquisition on the current set, cataloging of the file, and initialization of the next set.

6.2 QUANTITATIVE MEASUREMENTS AND QUALITY CONTROL

There are several important advantages gained by making concentration measurements by GC/MS. Quantitation with selected ion monitoring is widely used because it offers the ultimate in compound tunable selectivity, high sensitivity, precision and accuracy. Indeed the ultimate in quantitative accuracy is possible with SIM and a stable isotope labeled internal concentration calibration standard that is the same compound as the measured analyte. Because of the high selectivity of SIM, some reduction in sample purification is usually possible and this adds cost-effectiveness to the approach.

The use of data obtained by continuous repetitive measurement of spectra for quantitative analysis will clearly result in less sensitivity and perhaps lower accuracy and precision. However, the same tunable selectivity is available through the extracted ion current profile, and significant cost effectiveness is possible by use of the same data file for both qualitative and quantitative analyses.

There are a very large number of procedures for quantitative measurements with GC/MS. These range from standard manual measurements of peak heights or areas in the plotter display of the real time total ion chromatogram, to off line computations using digital data printed with the A option of the OUTP software (section 4.1). Measurement of peak heights is particularly effective with data acquired by the SIM programs CRTSIM or PLTSIM and displayed with the program PTSMPT. Manual methods are not discussed further in this section since these are generally understood by individuals experienced in chromatography work, and many authors have written on the subject.

In this section a number of terms used in quantitative analysis and quality control are defined, some guidelines for use of concentration calibration standards and quality control are presented, and instructions are given for the application of three computer programs to quantitative measurements. Some of the quality control concepts were designed for the analysis of water samples and these may not be applicable to other media. The computer programs are named QNTSET/QNTATE, QUANX (where x = a revision number,) and the IN option of MSSOUT. The MSSOUT software is described fully in section 4.2, and the user should consult this first to understand the basic operations of MSSOUT.

The following terms are defined to ensure that all readers understand the same meanings for several important terms.

External Concentration Calibration Standard (ECCS): A known amount of a pure compound that is measured separately (externally) from identically the same compound in a sample or sample extract. The measured detector response from the external standard is used to calibrate the concentration measurement of the same compound in the sample.

Internal Concentration Calibration Standard (ICCS): A known amount of a pure compound that is added to a sample or sample extract, but is not one of the compounds found in the sample. This compound may be labeled with an isotope or isotopes, but usually is not because it is not found in the sample. The measured detector response from the internal standard is used to calibrate the concentration measurements for other compounds in the sample or sample extract that are chemically different. This calibration is accomplished through independently measured response factors (RF) for each compound using the equation below. The independent measurements of RF's are made with an external standard that also contains the internal standard.

$$RF = \frac{\dfrac{\text{Area (X)}}{\text{Amount (X)}}}{\dfrac{\text{Area (S)}}{\text{Amount (S)}}}$$

where: Area (X) = the peak area of the compound in consistent units.

Amount (X) = the quantity of the compound injected in consistent units.

Area (S) = the peak area of the internal standard in consistent units.

Amount (S) = the quantity of internal standard injected in consistent units.

The final concentration of the unknown in ug/1 is computed with the equation below when an internal concentration calibration standard is employed. Areas and RF are as defined above.

$$C(X) = \frac{\text{Area (X)} * \text{Amount(S)}}{\text{Area(S)} * \text{RF} * \text{V}}$$

where: C(X) = the concentration of the unknown in the original sample in micrograms per liter

Amount(S) = the quantity of internal standard added to the concentrated extract or sample in micrograms

V = the volume of the original sample in liters.

Isotope Labeled Internal Concentration Calibration Standard (ILICCS): A known amount of a pure compound that contains a known proportion of an isotope or isotopes in an amount different from the naturally occurring amount. This compound is added to a sample or sample extract that contains the same compound with a natural distribution of isotopes. This is a special case of the internal concentration calibration standard which does not require the use of response factors. The presence of isotopes does not generally change a compound's response in a detector, or its extraction efficiency but the labeled and natural material may be distinguished with a mass spectrometer because of the differences in the masses of labeled ions. However, if the labeled internal standard were used to quantify chemically different compounds in the same sample, response factors would be required as described above.

Spike(s): A known amount of a pure compound that is added to the original environmental sample, and is the same compound as found in the original environmental sample. The measured percent recovery of the spike is taken as a measure of the accuracy of the total analytical method with the sample matrix, and when there is no change in volume due to the spike, it is calculated with the equation;

$$P = 100\,(O - X)/T$$

where: P = the percent recovery of the spike.

O = the measured value of the concentration of the analyte in the sample after the spike is added.

X = the measured value of the concentration of the analyte in the sample before the the spike is added.

T = the amount of the spike added expressed in terms of its concentration in the sample.

Surrogate Spike (SS): A known amount of a pure compound that is added to the original environmental sample, but is not one of the compounds found in the sample. The measured percent recovery of the surrogate spike is taken to be indicative of the recoveries of the compounds in the sample when actual spikes of each are not used. For surrogate spikes, the value of X in the equation above is zero.

Laboratory Control Standard (LCS): a sample that is prepared in the laboratory by dissolving a known amount of a pure compound in a known amount of clean water. The final concentration calculated from the known quantities is the true value of the standard. The measured percent recovery of the laboratory control standard is taken as a measure of the accuracy of the total analytical method independent of various sample matrices. The laboratory control standard must be carried through all the same sample processing that is used for the original environmental samples.

Check Standard (CS): A sample that is prepared in the laboratory by dissolving a known amount of a pure compound in a known amount of clean water or an organic solvent. The final concentration calculated from the known quantities is the true value of the standard. The check standard is not carried through all the same processing that is used for the environmental sample, but is often used to validate an existing concentration calibration standard file or calibration curve.

Duplicate (D): Two aliquots made in the laboratory of the same environmental sample. Each aliquot is treated exactly the same throughout the laboratory analytical method. The difference in the values of the duplicates is taken as a measure of the precision of the method.

Regardless of whether internal or external concentration calibration standards are employed, the standards and unknowns should be run under identical conditions in so far as this is possible. Therefore, GC temperature programming, integration times, and mass range in control runs, or integration times and number of masses monitored in SIM runs should be identical to cancel as much error as possible. No delay between scans should be utilized since more accurate integrals are obtained if the maximum number of spectra are measured for each peak. Good results are often far more dependent on the stability of the instrument sensitivity, baseline zero, and calibration than on the exactness of the quantitation program.

In general the use of an isotope labeled internal concentration calibration standard for each compound measured will be precluded because of the unavailability and/or high cost of such standards. Also the use of isotope labeled standards in this way requires fairly complex computations to account for background ions. These standards are best applied in specific situations for measurements of one or a few compounds.

For either internal or external concentration calibration standards it is necessary to measure the standards and, for internal standards, revalidate the response factors on every day that measurements are made of unknowns in samples. This is because some GC/MS system drift is inevitable and good results depend on system stability as much as any other factor. The daily measurement of an external standard and/or revalidation of response factors does not increase workload since this may be accomplished concurrent with a required daily measurement of an external standard to assure proper chromatographic resolution and system performance. It should also be established that concentrations are a linear function of peak areas for each compound measured, and the assumption of linearity should not extend beyond the established working range.

For sample extracts, external concentration calibration standards should be made in organic solvents; these standards are not carried through the total analytical method because extract losses could significantly affect the calibration process. For sample extracts, it is required that the internal concentration calibration standard or standards be placed in the concentrated sample extract no more than a few minutes before the measurement. The purpose of this is to protect the concentration standard from losses due to evaporation, adsorption, or chemical reaction.

With isotope labeled internal concentration calibration standards and solvent extracts, the standard is often added to the original sample to calibrate the total method, including the extraction efficiency. However this is a dangerous procedure since a relatively long contact time with the raw sample may result in losses of the calibrant, and cause serious errors in the measurement. This type of calibration should be attempted only if it is established that the internal standard is stable in the sample matrix. For solvent extracts, internal concentration calibration standards that are different compounds than the measured analyte should not be added to the original sample. This is because errors would result if the extraction efficiencies of the standard and measured analytes were different, but the response factors were determined without taking into account this difference.

The inert gas purge and trap procedure is an exception to the above because the internal concentration calibration standard must be added to the original sample. Therefore it is important that purging begin immediately after adding the internal standard, and even this may not preclude losses in the sample matrix. With the purge and trap procedure, external concentration calibration standards are made up in water. With this procedure the total method, including the purge efficiency, is calibrated with internal or external standards.

One of the disadvantages of an internal concentration calibration standard is that it must be added to a sample before the general concentration levels of the unknowns are known. Therefore the standard may be present in much greater or smaller quantity than the unknown. This may give rise to serious errors. External standardization has the advantage of allowing the concentrations of standards to be defined after making an estimate of the concentration levels of the unknowns. Therefore the standard/unknown concentration ratios can be adjusted to near unity which may lead to more accurate measurements.

Quantitative analysis by GC/MS requires another level of quality contol beyond concentration calibration. This is the day-to-day continuing evaluation of the precision and accuracy of the concentration measurements. The approach described here is oriented to the target compound approach and consists of two phases. In the first phase upper and lower control limits must be established for each method, target compound, and sample type using approximately fifteen laboratory control standards, fifteen spikes, and fifteen duplicates. Surrogate spikes should be included in all of these. The method of establishing control limits using percent recoveries and critical range factors is described in chapter six of the EPA Handbook for Analytical Quality Control in Water and Wastewater Laboratories.

Once these control limits are established for laboratory control standards, spikes, surrogate spikes, and duplicates, surrogate spikes should be added to every environmental sample to continuously monitor method and operator performance. As long as the surrogate spike recoveries are within the established control limits, it is assumed that the method is in general control. This must be confirmed approximately every 20–30 samples by measurements of a laboratory control standard, a spike, and a duplicate. If any of these control measurements exceed the control limits, all analyses should be suspended until the problem is corrected.

Of course blanks or reagent blanks must be used as recommended in Chapter 3 for qualitative as well as quantitative measurements. Measurements of a sample three or four times for determination of precision is not recommended. It is preferable to duplicate two samples than to measure one four times. Measured values of unknowns should not be corrected with the recovery data from a single spike, but the spike recovery data should be reported to the end user.

With these quidelines, it is feasible to use chemically reactive compounds, such as anthracene-d_{10}, for internal concentration calibration. However, better choices are available including many brominated and fluorinated compounds that are not likely to be as chemically reactive and variable in composition. Several possibilities exist including 1,4-dibromobenzene, a tribromobenzene, pentafluorobromobenzene, or perfluorobenzene.

Although there is very little published data, quantitative analysis with continuous repetitive measurement of spectra and a single external concentration calibration standard in one laboratory gave an average bias of –30% for laboratory control standards in the 0.1 – 10ug/1 range.

All of the computer programs described below are compatible with the use of internal concentration calibration standards. This is accomplished by entering the same data file name for the standard and the unknown. More specific instructions for internal standards are included in the description of each program. The QUANX program is particularly well adapted to internal standards since it contains an option to enter response factors.

The programs QNTATE and QUANX quantify an unknown by a simple linear extrapolation from the peak area measurement of a single standard.

Therefore it is desirable for the concentration of the standard to be reasonably close to the concentration of the unknown. A factor of 3 or 4 is considered reasonable but linearity should be established as described above. The program MSSOUT does not do this computation, but merely prints out the integrated peak areas for various peaks from various files. The user must make the final calculation and may include a response factor if desired.

In the sections that describe the individual programs, the examples were generated from the same two data files and the results obtained give a direct comparison of the three programs. The data files were obtained from 50ng and 100ng injections of DFTPP and are representative of clean, well resolved, well shaped peaks. Therefore the results obtained are probably the best that can be expected from the programs.

MEASUREMENTS WITH QNTSET/QNTATE

This pair of programs has several advantages. They may be used with data files acquired under the standard control mode, IFSS, or one of the specialized SIM programs CRTSIM or PLTSIM. QNTSET is used to create a file of peak area data using information from a data file generated by a concentration standard; and QNTATE is used to relate standard and unknown peak areas and print out concentrations in nanograms. A disadvantage of this pair of programs is that they cannot be used in the extracted ion current mode, i.e., if the standards or unknowns were measured with the control or IFSS mode using a range of mass units, all the abundances measured at all the masses are used in the quantitation. There is no provision to extract abundances at specific masses and use only these for quantitation. However, if data was acquired with control, IFSS, CRTSIM, or PLTSIM at a specific mass or masses, clearly only the specific mass or masses will be used in the quantitation. Peak areas in both QNTSET and QNTATE are computed with the baseline subtracted. The user specifies the spectral limits of the peak and the program then assumes that the baseline connects the first spectrum before the starting value and the first spectrum after the terminating value.

Sample dialogue for QNTSET is as follows:

```
SELECT MODE:  QNTSET
STANDARDS FILE NAME?  STAN
SPECTRUM FILE?  REF100
FIRST SPECTRUM?  68
SECOND SPECTRUM?  75
SPECTRUM NUMBER OF MAXIMUM?  71
NAME OF COMPOUND?  DFTPP
```

```
NANOGRAMS OF COMPOUND?  100
SAME FILE?  N
NEW FILE?  N
```

The STANDARDS FILE NAME is not a data file from the GC/MS of the standards, but a new file to be created by QNTSET that will contain peak area data from the standards. Therefore a new file must be named by the user. The SPECTRUM FILE is the name of the data file from the GC/MS of the standards. The FIRST SPECTRUM is the first spectrum number of a GC peak, the SECOND SPECTRUM is the final spectrum number of the same GC peak, and the SPECTRUM NUMBER OF MAXIMUM is the spectrum number from the same GC peak where the total ion abundance reaches a maximum. The NAME OF THE COMPOUND may be entered by the user and the quantity of material expressed in nanograms injected. This must be expressed as an integer number, e.g., if one microgram was injected, the user should enter 1000.

The SAME FILE? prompt should be answered positively if the user wishes to enter peak area data for another peak from the same GC/MS data file into the same peak area file. The NEW FILE? prompt should be answered positively if the user wishes to enter peak area data for another peak from a different GC/MS data file into the same peak area file. A negative response to this last prompt causes the peak area file to be closed and saved and the user is returned to the system prompt. A response of CTRL/L to any prompt causes the program to abort, and all peak area data will be lost.

The QNTATE program requires the presence of the peak area file generated by QNTSET. Sample dialogue is as follows:

```
SELECT MODE:  QNTATE
SPECTRUM FILE?  REF50
FIRST SPECTRUM?  68
SECOND SPECTRUM?  75
SPECTRUM NUMBER OF MAXIMUM?  71
STANDARDS FILE NAME?  STAN
PEAK NUMBER:  1
DFTPP 54 NANOGRAMS
SAME FILE?  N
NEW FILE?  N
```

The SPECTRUM FILE is the GC/MS data file containing a peak or peaks to be quantified. The FIRST SPECTRUM, SECOND SPECTRUM, and SPECTRUM NUMBER OF MAXUMUM prompts have the same meaning as described under QNTSET. The STANDARDS FILE NAME is the peak area file generated by QNTSET. The response to the PEAK NUMBER prompt is the number of the entry in the peak area file which is to be used to quantify the unknown peak. For example, if the user has created a peak area file containing three entries such as the following, and wants to quantify a DDD peak, the reply to PEAK NUMBER would be 2.

1. DDT 50 nanograms
2. DDD 100 nanograms
3. Malathion 80 nanograms

The program then prints the name of the compound stored in the peak area file, and the computed number of nanograms in the unknown. In the example 50 nanograms of DFTPP were taken and 54 nanograms found using a 100 nanogram standard.

The SAME FILE prompt should be answered positively if the user wishes to quantify another peak from the same GC/MS data file. The NEW FILE prompt should be answered positively if the user wishes to quantify a peak from a different GC/MS data file.

MEASUREMENTS WITH QUAN3

This program works with data files acquired under the control or IFSS modes or with the specialized SIM programs CRTSIM and PLTSIM. The program is somewhat simpler to use than QNTSET and QNTATE. Furthermore, it has the major advantage of operating in the extracted ion current mode, i.e., if the standards or unknowns were measured with the control or IFSS mode using a range of mass units, QUAN3 may be used to quantify with a single mass range shorter than the measured mass range, or a single mass unit, or even a single fractional mass unit. The program works by summing for all spectra between given limits (inclusive) the abundances of all masses within the given range (inclusive.) The baseline is subtracted during the integration and the user may enter a response factor which is a weighting factor for the measurement. Sample dialogue for the program is as follows:

```
SELECT MODE: QUAN3
STANDARD FILE NAME: REF100
SPECTRUM NUMBERS: 68–75
MASS RANGE: 442
SUBTRACT BASELINE: Y
AMOUNT = 100
```

```
ABSOLUTE INTENSITY = + 124928.000

UNKNOWN FILE NAME:  REF50
SPECTRUM NUMBERS:  68–75
MASS RANGE:  442
SUBTRACT BASELINE:  Y
RESPONSE FACTOR:  1

ABSOLUTE INTENSITY = + 67328.000

AMOUNT = + 53.893

UNKNOWN FILE NAME:
```

The data files used to illustrate this dialogue are the same as those used to illustrate QNTSET/QNTATE and MSSOUT. The STANDARD FILE NAME is the data file generated by GC/MS of the standards, in this case 100 nanograms of DFTPP. Quantitation was accomplished using only molecular ion data although the original data acquisition was over the 33–450 amu mass range. Using a response factor of 1, the program found 53.893 nanograms of DFTPP in the 50 nanogram injection. The unknown amount is given in the same units as the standard amount.

MEASUREMENTS WITH MSSOUT

The integrate (IN) command of MSSOUT may be used to integrate peak areas after the CM and EM commands are issued according to the standard rules. The CM and EM commands are described in section 4.2. The IN command has the same advantage of QUAN3 in that quantitation may be accomplished using individual masses although abundance data from the entire mass range was acquired originally. However, MSSOUT has no apparent capability to relate standard and unknown peak areas. Therefore this final calculation, including the response factor, must be accomplished by the user with a calculator or some other manual procedure. Sample dialogue for the IN option of MSSOUT using the same two DFTPP data files is as follows:

```
SELECT MODE:  MSSOUT

RUN NAME: REF100
DFTPP 100NG
*CM
ENTER MASSES TO COLLECT. MAX = 47
442
```

```
*EM
ENTER MASSES TO PLOT
442

*IN
START = 68
  END = 75

MASS  SUM
442   126148

*NA

RUN NAME:  REF50
DFTPP 50NG
*CM
ENTER MASSES TO COLLECT. MAX = 47
442

*EM
ENTER MASSES TO PLOT
442

*IN
START = 68
  END = 75

MASS SUM
442   68360

*
```

As in plotting an EICP with MSSOUT, the specific mass to be used in the IN command must be collected with the CM command. The START and END prompts produced by the IN command request the starting and ending spectrum numbers of the peak. In place of entering the spectrum numbers from the keyboard, the user may touch the space bar which displays the cross hair cursor. Moving the cursor to the left and right sides of the peak, and pressing the space bar after each, causes input of the spectrum numbers. The NA command changes the name of the data file from the standard file to the unknown file. Final calculation using the formula described above gave 54.2 nanograms for the 50 nanogram injection.

Peak areas for more than one ion across a single peak are obtained as shown below; in this case EM is required to define the sequence of events for the IN command.

```
*EM
ENTER MASSES TO PLOT
198,442

*IN
START = 67
  END = 79

MASS SUM
198  100992
442   79752
```

6.3 OPEN TUBULAR COLUMNS

The purpose of this section is to review the current advantages, disadvantages, and applications of open tubular columns in environmental monitoring with GC/MS. It is clearly beyond the scope of this manual to review in depth the entire field of open tubular columns which is a very active area of current research.

The primary application for open tubular columns in environmental monitoring is in the analysis of very complex samples such as industrial and other waste effluents. Another application is in the analysis of samples that contain difficult to separate but environmentally significant components such as both polychlorinated biphenyls and chlorinated benzodioxanes. However the user must recognize that the use of high performance, high resolution open tubular columns introduces trade-offs in analytical operations. Greatly improved resolution invariably requires significantly more control of operational variables. Therefore there is an important need to understand the additional costs required to produce the improved performance, and to apply the high performance columns only in situations that truly require them. For the majority of problems conventional packed columns currently offer the best combination of performance and operational convenience. Also the user should be aware that the quantitative capabilities of open tubular columns are not well established.

Another factor that should be recognized is that there is a far greater need for high resolution performance with conventional GC detectors than with a mass spectrometric detector. With conventional detectors, the retention index is the only output useful for qualitative analyses. Therefore it is vital to acquire accurate and precise retention data, and it is not surprising to find that the leading proponents of open tubular columns are users of conventional detectors. With the mass spectrometric detector and the ability to acquire several mass spectra across many GC peaks, unresolved components may often be clearly recognized, identified, and even quantitated using characteristic ions from each component. Similarly with selected ion monitoring high resolution often may be of no real value. Nevetheless with complex mixtures of compounds with complex spectra, open tubular columns offer outstanding capabilities for GC/MS application and may be the method of choice.

OPERATING VARIABLES

As previously indicated, it is not the intention of this section to review comprehensively all possible operational variables. However the ideas presented seem to represent a consensus of the important viewpoints of the most successful and experienced users of open tubular columns.

By all means do not attempt to coat open tubular columns unless you have a great deal of experience or the time to acquire it. Commercial columns of high quality are available from several sources, and the time required to master and control the coating technique is much greater than the cost of the commercially available columns. Glass columns are strongly recommended because they are chemically inert with respect to most compounds.

Splitless injection is strongly recommended. The method of stream splitting should be abandoned because of significantly reduced sensitivity due to the discarding of a large fraction of the injected sample, and the uncertainty of reproducing the split ratio. A particularly important variable with splitless injection is the solvent. In general a solvent should be selected so the initial column temperature may be set at least 20C below the boiling point of the solvent. This will permit the operation of the so called solvent effect which acts to concentrate the sample components into very narrow bands giving maximum peak resolution. The choice of solvent will usually be constrained by the capabilities of the GC to achieve ambient or sub ambient column oven temperatures. If the GC equipment in use does not have low temperature capability, the potential for achieving optimum peak resolution with splitless injection and the most desirable solvent will be sharply reduced. For example, methylene chloride is strongly recommended in chapter 3 for low boiling solvent extraction. Methylene chloride is also an excellent solvent for use with open tubular columns. However the solvent effect would cause an undesirable peak broadening effect with methylene chloride unless a satisfactory initial column temperature could be obtained. If sub ambient temperature capability is not available, and high performance is needed, a possible alternative is replacement of the low boiling solvent with a higher boiling one. This of course may not be practical because of solubility or other considerations, adds to the complexity of the procedure, and the higher boiling solvent could contribute undesirable background to the sample.

Two of the most critical factors which operate to reduce column lifetime to unacceptable levels are repeated violations of the high temperature limit of the column, and the injection of samples containing water. These conditions should be avoided. With splitless injection samples of the order of 1–2 microliters may be injected into the system. Columns prepared by commercial suppliers using the newer methods of manufacture can withstand routine direct injections of several microliters of pentane, hexane, methylene chloride, or carbon disulfide for months without damage. Some solvents, such as methanol, are not recommended, but the manufacturer's instructions should be carefully noted when choosing solvents.

The application of open tubular columns has several implications for all data systems. Very sharp GC peaks require scan speeds that are relatively fast, and if all mass spectra are saved on a computerized storage device, sufficient space must be available to store the potentially large number of spectra generated during a long elution period. The PDP-8 data system has several features that facilitate the use of open tubular columns, but reasonable caution is required for this and any other data system that was not specifically designed for this application.

In section 2.7 under FAST SCAN OPTION and the following paragraphs, the basic sweep speed considerations of the PDP-8 data system are explained. Insertion of the fast scan option hardware causes the settling time for each mass set voltage to be reduced from 2.3 msec to 1 msec, with some improvement in scan speed. Another method of improving sweep speeds is to apply the IFSS method of data acquistion. The following dialogue shows operating parameters that give a 2.5 sec sweep with a 2.3 msec settling time. Use of a 1 msec base integration time reduces the sweep time to about 2 sec.

```
SELECT MODE: IFSS
CALIBRATE?: N
TITLE: OPEN TUBULAR TEST
CALIBRATION FILE NAME: CAL
FILE NAME: SAMPLE
MASS RANGE: 33-450
SAMPLES/AMU: 1
MAX RPT COUNT: 5
```

```
BASE INTEGRATION TIME:  2
RPT COUNT BEFORE CHECKING LOWER THRESHOLD:  2
LOWER THRESHOLD:  3
UPPER THRESHOLD:  1
RT ON CRT?:  N
RT GC ATTEN:  5
MS RANGE SETTING?:  H
MAX RUN TIME:  30
DELAY BETWEEN SCANS (SECS.)?
```

DATA

Software exists to cause a data file to begin on the first disk drive (unit 0) of a two disk system, and conclude on the second disk drive (unit 1) if insufficient space is available on unit 0. The files required are TWIN and EXMOD2, and bit 11 of the hardware status word at location 2060 in INIT$$ must be a one (see section 7.1).

6.4 CHEMICAL IONIZATION(CI)

The purpose of this section is to review the current advantages, disadvantages, and applications of chemical ionization (CI) in environmental monitoring. It is not a purpose to review the entire field of CI, which is very broad in scope and a very active area for current research.

The primary application of chemical ionization in environmental monitoring is to obtain the additional information necessary to accurately identify compounds that are not reliably identified through their electron ionization (EI) spectra. Most often additional information will be required in situations where a molecular ion is not observed in the EI spectrum, and where an authentic standard is not available for acquisition of other supporting information such as a GC retention time. The problem of quality assurance in compound identification is discussed further in section 5.3.

Another important application of CI is in selected ion monitoring where ions formed in CI processes are sometimes more selective and abundant than ions in the EI spectrum of the same compound. An advantage of CI is that significant diagnostic fragment ions are sometimes present that are not observed in the EI spectrum. This advantage would be of value to those engaged in the interpretation of spectra of unknown compounds whose EI spectra do not appear in mass spectral libraries. A potential future application is negative ion CI. Negative ions have been shown to be generated by some compounds in much greater abundance under CI conditions than under EI conditions. However, most current instruments cannot observe negative ions without major modifications, and the significance of this technique is not yet fully demonstrated.

At the present time most types of chemical ionization cannot stand alone as the primary method of ionization in broad spectrum organic analyses as defined in this manual. There are several reasons for this situation. First many compounds give, with the popular reagent gases, CI spectra consisting mainly of molecular ions and molecular ions containing a proton or some hydrocarbon fragment. Therefore these spectra are less useful for identifications than the corresponding EI spectra which often contain many structurally significant fragment ions. Secondly, although many compounds show fragmentation in their CI spectra, CI spectral libraries have not yet been developed. Therefore, interpretations of spectra are required, and this is a relatively slow process compared to empirical spectrum matching.

In addition to the problem of the sometimes confusing M+29, M+41, and other extraneous ions in some types of CI spectra, there are problems associated with the large number of operational variables of CI. These variables sometimes lead to significantly different CI spectra produced from the same compound by different laboratories and instruments. This makes interlaboratory comparisons difficult and discourages spectral library development. The principal variables appear to be reagent gas, ion source design, ion source temperature, and ion source pressure. Other variables such as certain ion source potentials, electron energy, etc. may also be important. One approach to generating comparable spectra from different instruments in different laboratories is the use of a standard reference compound to calibrate the experimental variables. Of course this should be used in connection with temperature and pressure gauges and other operating parameter controls. Unfortunately some equipment sold commercially does not include essential controls. Also temperature, pressure, etc. gauges are not always reliable indicators because these devices depend on transducers that are subject to variable-output after aging or contamination.

METHANE CHEMICAL IONIZATION REFERENCE COMPOUND

The EI GC/MS reference compound decafluorotriphenylphosphine (DFTPP) is particularly well suited as a methane CI reference compound. The methane CI spectrum of DFTPP displays, in addition to an abundant M+1 ion at mass 443, several major fragment ions that are of very low abundance in the EI spectrum. Therefore the CI spectrum is an excellent indicator of the presence of correct CI conditions. Figure 6.5 shows the dependence of the fragmentation pattern of DFTPP (methane CI) on the ion source temperature. At temperatures below 150C the ions at masses 423 (loss of F) and 365 (loss of a phenyl group) are the most abundant in the spectrum. In the corresponding EI spectrum both of these ions are below 5% relative abundance. As the temperature of the ion source in the CI experiment was increased, thermally induced fragmentation increases until, at temperatures above 200C, the base peak is mass 169 and very little M+1 ion was observed. The methane CI spectrum of DFTPP at an ion source temperature of 90C as displayed in Figure 6.5 is proposed as a reference spectrum for methane CI measurements.

The methane CI of DFTPP is much less sensitive to ion source pressure in the range commonly used for CI measurements. However overall sensitivity is a significant function of ion source pressure.

PRINCIPLES OF CHEMICAL IONIZATION

In chemical ionization electrons are used to ionize a reagent gas which subsequently ionizes sample molecules by chemical reactions, i.e. proton transfer, hydride abstraction, ion attachment, etc. Chemical ionization is usually a much less energetic process than electron ionization. Electron ionization generates molecular ions with an internal energy of 500 to 1000 kcal/mole, and in order to release such energy, these ions may undergo rearrangements and fragmentations. Often the most important piece of information in a mass spectrum, the mass of the molecular ion, is lost, and only fragment ions will be apparent. In contrast, initial ions formed by a chemical ionization process usually have an internal energy of only 50 to 100 kcal/mole, which means that CI spectra usually exhibit intense ions in the molecular weight region. Chemical ionization has been shown to offer sensitivity at least as good as electron ionization, and offers the advantage of allowing characterization of the sample's chemical reactivity through the choice of reagent gases. While methane has been the most widely used reagent gas, a number of others, singly and in combination, have been successfully used to produce CI mass spectra. Popular reagent gases include hydrogen, helium, argon-water, ammonia, and nitric oxide.

When methane is bombarded with high energy electrons, primary reaction 1 takes place.

$$CH_4 + e^- \longrightarrow CH_4^+ + CH_3^+ + CH_2^+ + CH^+ + H_2^+ + H^+ + 2e^- \quad (1)$$

If the gas is present in a confined chamber at a pressure of approximately one mm of Hg, i.e. if it is in a CI mass spectrometer source, the following secondary reactions can take place (2 through 5):

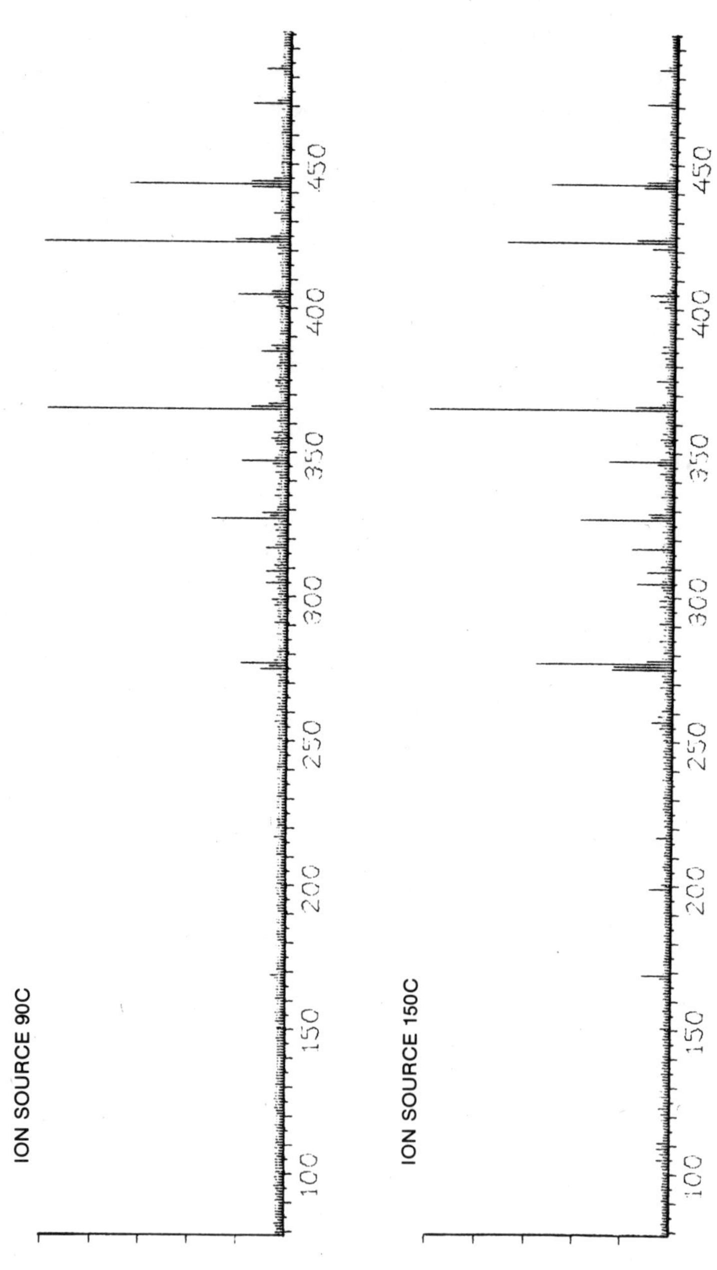
ION SOURCE 90C
ION SOURCE 150C
100
150
200
250
300
350
400
450

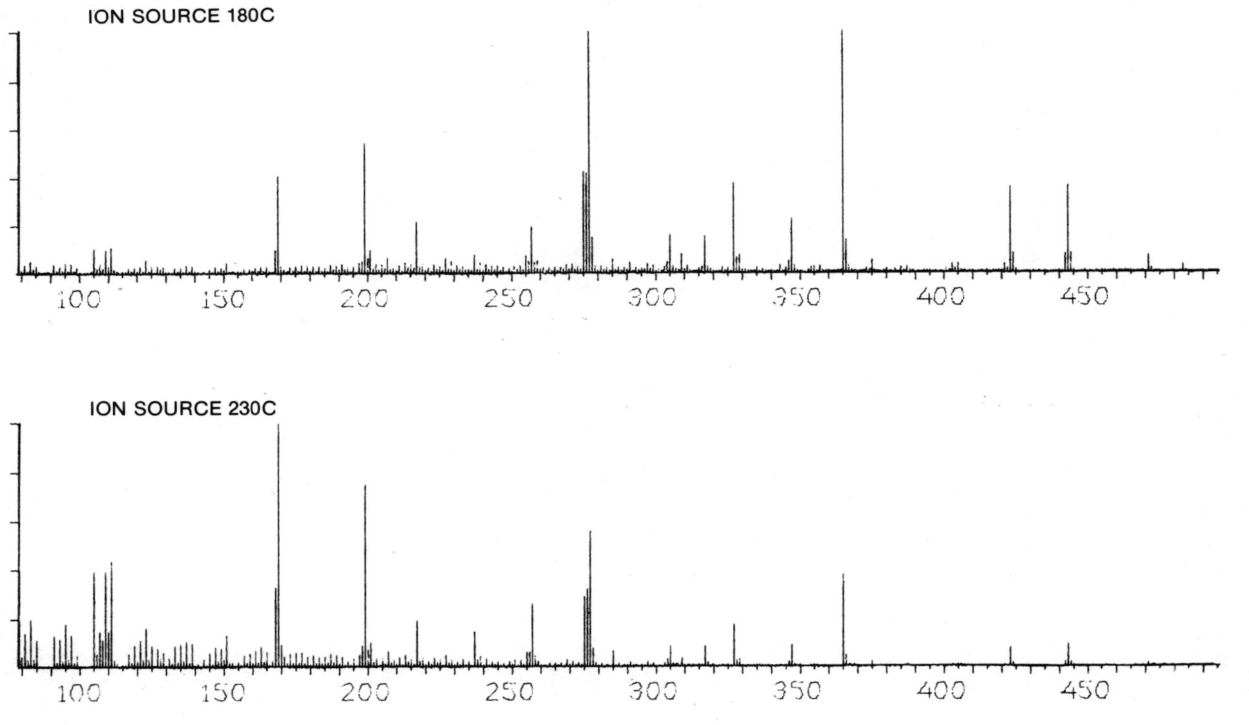

Figure 6.5 Chemical ionization (methane) mass spectra of DFTPP as a function of ion source temperature on the Finnigan model 4000 GC/MS System.

$$CH_4^+ + CH_4 \longrightarrow CH_5^+ + CH_3 \quad (2)$$

$$CH_3^+ + CH_4 \longrightarrow C_2H_5^+ + H_2 \quad (3)$$

$$CH_2^+ + CH_4 \longrightarrow C_2H_3^+ + H_2 \quad (4)$$

$$C_2H_3^+ + CH_4 \longrightarrow C_3H_5^+ + H_2 \quad (5)$$

In a normal CI source, at a pressure of one mm, CH_5^+ and $C_2H_5^+$ will comprise about 90% of the total ionization, while $C_3H_5^+$ will be about 5%. If a sample molecule, MH, is present at a pressure much less than the reagent gas, it can react with the above ions in the following manner (Reactions 6 through 11):

$$CH_5^+ + MH \longrightarrow CH_4 + (MH_2)^+ \quad \text{(proton transfer)} \quad (6)$$

$$CH_5^+ + MH \longrightarrow CH_4 + H_2 + (M)^+ \quad \text{(hydride abstraction)} \quad (7)$$

$$C_2H_5^+ + MH \longrightarrow (MH + C_2H_5)^+ \quad \text{(ion capture)} \quad (8)$$

$$C_2H_5^+ + MH \longrightarrow C_2H_4 + (MH_2)^+ \quad \text{(proton transfer)} \quad (9)$$

$$C_3H_5^+ + MH \longrightarrow (MH + C_3H_5)^+ \quad \text{(ion capture)} \quad (10)$$

$$C_3H_5^+ + MH \longrightarrow C_3H_6 + (M)^+ \quad \text{(hydride abstraction)} \quad (11)$$

Figure 6.6 shows a comparison of the EI and CI mass spectra of octadecane. The EI spectrum shows a base peak at mass 57. The CI spectrum shows the characteristic ion at mass 253 (M-1) formed according to equations 7 and 11.

There are usually two options for handling the carrier gas in CI. The carrier gas may be used as the chemical ionization reagent gas, and no enrichment device is required. This has the advantage of improved sensitivity since all of the chromatographic effluent enters the spectrometer. It has the disadvantage of requiring the use of unusual carrier gases and perhaps changing chromatographic behavior. Alternatively the usual GC carrier gases may be used with an enrichment device, and then the CI reagent gas introduced into the GC inlet line just prior to the effluent entering the ion source. With this method some sample will be lost, but the customary GC carrier gases may be retained with the considerable flexibility of adding a variety of reagent gases. The latter approach is necessary with open tubular columns since the low flow rates used are inadequate to give appropriate reagent gas ion source pressures. A modern, well designed chemical ionization mass spectrometer should be equipped for all of the above options, and a system for rapid switching among them.

6.5 ACCURATE MASS MEASUREMENTS

The purpose of this section is to briefly review the concept of accurate mass measurements and their application to compound identification. As with several other sections in this chapter, a comprehensive review of the subject is beyond the scope of this manual, and the user is directed for more information to several excellent books referenced in the Bibliography. Although only a small fraction of the mass spectrometers manufactured have accurate mass measuring capability, it is important that the user of a conventional GC/MS system recognize the existence of this capability. For certain samples of particular significance the most important decision may be to make arrangements for an accurate mass measurement of one or more ions.

Accurate mass measurements are defined as measurements made to 0.01 amu or better and are important because this degree of accuracy may allow an accurate determination of the atomic composition of the ion. This information is a powerful aid to the identification of unknowns, particularly in situations where a matching or similar spectrum is not available in a mass spectral library. For example, both acetone and *n*–butane exhibit a molecular ion at mass 58. However, because of the mass defect of the constituent atoms of each ion, i.e., the deviation of the actual atomic mass from the nominal integer number, the acetone ion has an accurate mass of 58.0417 while the isobutane ion has a mass of 58.0780. Therefore, these two ions differ in mass by 36.3 millimass units (mmu). Both have a positive mass defect, that is, their mass defects are in excess of the nominal integer value. Mass defects can also be negative, in which case the accurate mass is less than the integer mass number. In order to distinguish between these two adjacent mass spectral peaks, they must be separated or resolved. One common definition of resolution states that two peaks are resolved when the valley between them is ten percent of the overall peak height. Resolution is mathematically defined as M/deltaM, where M is the nominal mass of the two ions to be separated, and deltaM is the difference between them. Using the example of mass 58 ions of acetone and *n*–butane, a resolution of 1600 would be needed to resolve them. The calculation is shown below.

$$\text{Resolution} = \frac{\text{M}}{\text{deltaM}} = \frac{58}{.0363} = 1600$$

This is considered moderate resolution. Resolution in excess of 8000 is considered high since that amount is usually necessary to resolve most mass

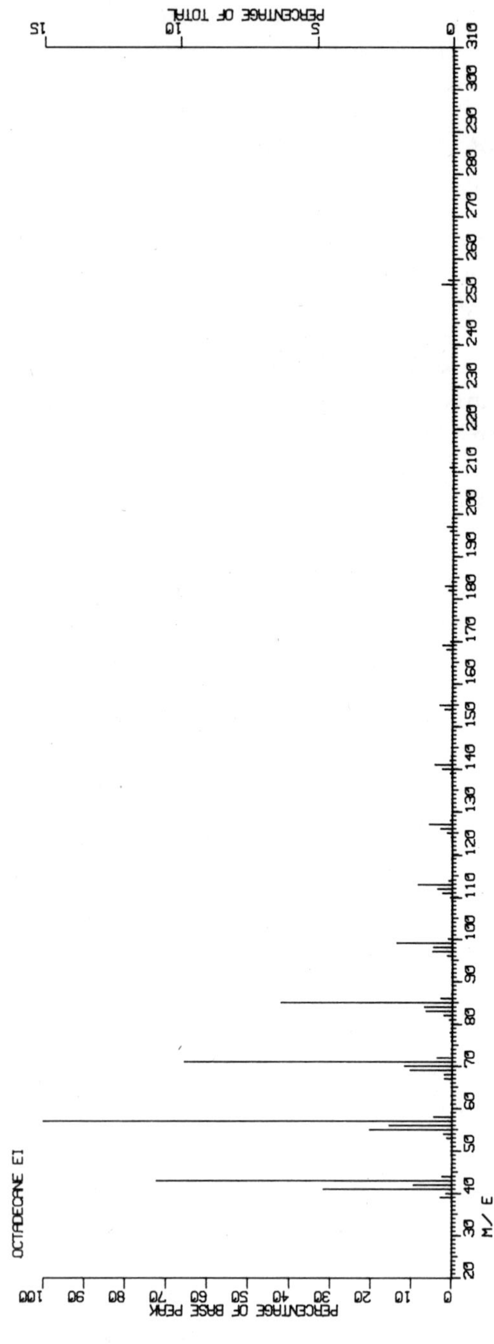
OCTADECANE EI
PERCENTAGE OF BASE PEAK
PERCENTAGE OF TOTAL
M/E

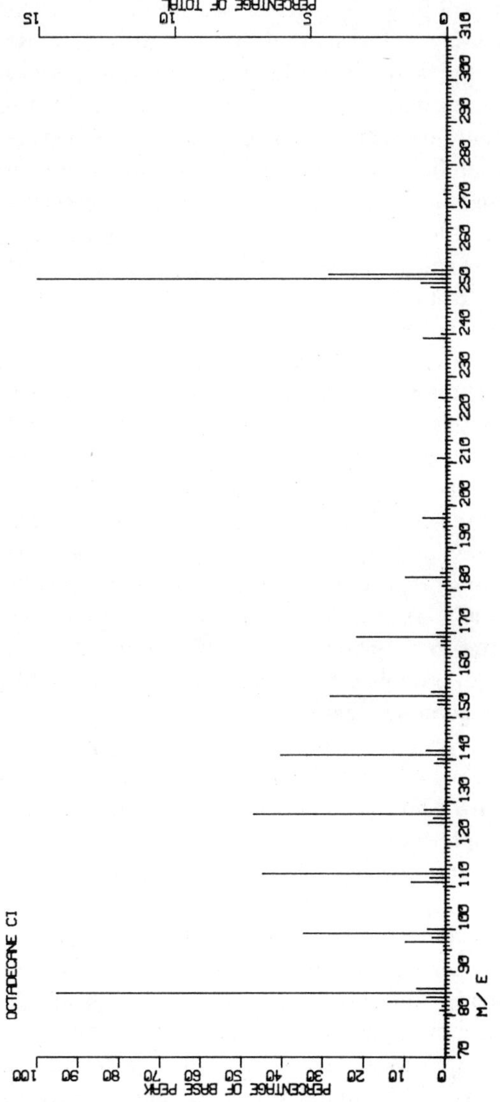

Figure 6.6 Comparison of the EI and CI mass spectra of octadecane.

doublets. In order to achieve such resolution, double focusing of the ion beam is usually necessary. A magnet focuses only on the basis of mass, but an electrostatic analyzer, i.e. a set of curved plates with a voltage impressed on them, will focus ions on the basis of their kinetic energy. Adding such extra focusing reduces the overall quantity of ions traversing the mass spectrometer, and thus reduces overall sensitivity. To overcome such a reduction, the mass range is usually scanned at a slow rate. To minimize the effects from slow scanning and decreased sensitivity, one should only use as much resolution as necessary to perform the required analysis, always keeping in mind that the accuracy of a mass measurement is independent of resolution as long as any mass doublets are separated.

Accurate masses of ions can be determined by several methods, but the two most common are peak matching and computer acquisition. Peak matching involves a comparison of the unknown mass spectral peak wih a known ion peak, usually from a standard material such as perfluorokerosene (PFK). By switching the accelerating voltage rapidly such that the unknown and known ions are alternately focused on the collector, one can overlay or match the two ions, measure the ratio of their masses, and therefore determine the mass of the unknown ion. The computer acquisition method of determining accurate masses depends upon a computer finding two relatively close reference masses, and then interpolating between them to identify the exact mass of an unknown ion.

Once the mass of an ion has been accurately measured, it is a relatively easy matter to determine what combination of atoms will add up to that number. Several tables have been published which list accurate masses and the corresponding composition formulas. However the most effective approach is to use a computer program to find all possible compositions whose masses are within the probable error of the accurately measured mass. Listed below is an interactive program written in the Basic language that finds all possible carbon, hydrogen, nitrogen, and oxygen compositions that fall within the range of a measured mass plus or minus the probable error. The program is constrained further by a user imposed limitation on the number of oxygens or nitrogens it may use in any given formula. This constraint was inserted to rule out a large number of formulas containing unreasonable numbers of these elements. It is usually possible to guess at the maximum number of oxygens and nitrogens, run the program, and if too few or too many formulas are found, change the guess and run it again. Also no formulas will be found which contain an extremely small number of hydrogens relative to the number of carbons. Atoms other than N or O could be included in the program by modifying it, or by substituting their masses for N or O in the program.

```
0002 PRINT "WHAT IS YOUR PRECISE MASS, INTEGER
       AND DECIMAL PARTS?"
0004 INPUT M
0006 PRINT "WHAT IS YOUR ERROR, E.G., .002,?"
0008 INPUT E
0010 PRINT "WHAT IS THE MAXIMUM NUMBER OF
       OXYGENS TO BE CONSIDERED?"
0012 INPUT O
0014 PRINT "WHAT IS THE MAXIMUM NUMBER OF
       NITROGENS TO BE CONSIDERED?"
0016 INPUT N
0018 LET A=M/12
0020 LET A=INT (A)
0022 LET L=INT (M)
0024 FOR J=0 TO O
0026 FOR K=0 TO N
0028 LET C=A
0030 LET B=L-C*12-J*16-K*14
0032 IF B<C/2 THEN GOTO 0045
0034 LET H=B*1.00782522+C*12+J*15.994915+K*14.003074
0036 IF M+E<H THEN GOTO 0045
0038 IF H<M-E THEN GOTO 0045
0040 PRINT "C";C;"H";B;"N";K;"O";J,H,H-M
0045 LET C=C-1
0047 IF B+12<=2*C+2+K THEN GOTO 0030
0049 NEXT K
0051 NEXT J
0053 END
```

To illustrate the effect of the precision of the mass measurement on the number of compositions found, the program was run with the input parameters and results shown in Table 6.1. The first measurement at about the accuracy possible with a conventional quadrupole GC/MS system leads to 37 possible formulas ranging from $C_{18}H_{20}$ to $C_8H_{18}N_3O_5$. This is far too many for a practical analysis of the data. The second measurement gave the following compositions and masses:

$C_{15}H_{14}N_3$	236.11878
$C_{16}H_{14}NO$	236.10754
$C_{12}H_{16}N_2O_3$	236.1161
$C_{11}H_{14}N_3O_3$	236.10352
$C_{13}H_{16}O_4$	236.10486

If the user can clearly establish that the compound contains no nitrogen, this measurement would give an unequivocal composition. The final measurement, which was made to ± 8 ppm, gives only one possible composition, $C_{16}H_{14}NO$. This level of precision is well within the capability of many modern mass spectrometer systems.

In conclusion, it should be stated that accurate mass measurements are relatively expensive, require more time per analysis, and are not as sensitive as conventional measurements. But they do provide the analyst with information that is unavailable from other techniques, and allow one to make identifications and characterize compounds with much more certainty.

Table 6.1 The Effect of the Accuracy of Mass Measurements and Atom Constraints on the Number of Acceptable Compositions

Measured Mass	Probable Error ±	Maximum Number of Nitrogens	Maximum Number of Oxygens	Number of Acceptable Compositions
236.1095	0.1	3	5	37
236.1095	0.01	3	5	5
236.1095	0.002	3	5	1

CHAPTER 7
AUXILIARY SOFTWARE

A number of computer programs are used directly in the acquisition and output of data from the computerized GC/MS system. The use of these programs is described, with specific examples, in the chapters on GC/MS operations and Quality Assurance (Chapter 2), Data Output (Chapter 4), Compound Identification (Chapter 5), and Advanced Analytical Techniques (Chapter 6). There is another group of computer programs which are used to evaluate the performance of certain hardware components of the data acquisition and control system. The use of these programs is described, with specific examples, in the chapter on Preventive Maintenance (Chapter 8) and Trouble Shooting (Chapter 9). Finally there is a miscellaneous group of computer programs that are useful in working with the GC/MS system, but do not fit into any of the above categories. The purpose of this chapter is to describe the application of these with specific examples. Some of these programs operate under the GC/MS real time operating system, and some of them are best used only with an auxiliary operating system. The choice of an auxiliary operating system is of some importance as several systems are available.

The OS-8 operating system requires 8K of core memory and has the advantage of being supported and maintained by the Digital Equipment Corporation. It is very widely used by thousands of installations across the country, and a great deal of utility software is available for use with it. However, it would be impossible to switch all the mass spectrometer programs over to OS-8 because it is not a real time operating system.

The X-System is an older utility system that requires only 4K of core memory. It was developed at Stanford University and System Industries acquired it from the Digital Equipment Corporation Users Society (DECUS). As far as is known it is not supported or maintained by anyone, a large number of bugs exist in some of the programs on some versions, and the kind of software usable with it is extremely limited.

For the reasons given above the OS-8 system is strongly recommended as the standard PDP-8 auxiliary operating system. The X-system will not be discussed in this manual, and any user who wishes information about it may acquire limited documentation from DECUS (DECUS No. 8–64a) or System Industries.

7.1 REAL TIME OPERATING SYSTEM

All the programs discussed in this section are found on the standard system disk and are used under that operating system. Additional programs may be loaded onto and run from the real time operating system. Instructions for loading these are given in section 7.2.

TAPE INITIALIZATION PROGRAMS

These programs are required to initialize a formatted Dectape or industry standard 9 track magnetic tape for use with programs or data files. The program NEWTAP is used to initialize Dectapes with either type controller (see section 9.2):

```
SELECT MODE:  NEWTAP
MOUNT NEW DECTAPE ON UNIT 1
(TC08 controller)
MOUNT NEW DECTAPE ON UNIT 0
(TD8E controller)

SELECT MODE:  LIST
TAPE?:  Y

NAME BLOCK

ENTRIES LEFT  =174
BLOCKS LEFT  =  5525
NEXT AVAILABLE BLOCK  =  15
```

If you have available an industry standard 9-track magnetic tape with the Datum controller, the corresponding program is NEWMAG. The Dectape unit must be write enabled and an industry standard magnetic tape must have a write ring inserted.

FILE COPY, DUPLICATION, AND SAVE PROGRAMS

Program and data files may be copied from one storage device to another (Dectape, disk, or industry standard magnetic tape) with the COPY program. Sample dialogue for disk to disk transfer is as follows:

```
SELECT MODE: COPY
FILE NAME: FILE TO BE COPIED
RENAME? FILE MAY BE RENAMED
FROM DISK? Y
UNIT NO.: 0
ONTO DISK? Y
UNIT NO.: 1
```

If a transfer to or from Dectape is desired, a N response is made to the FROM DISK? prompt or the ONTO DISK? prompt. This causes the program to enter the Dectape dialogue. The user must be aware that there are two different programs named COPY. One version copies files among disk units 0–3 and TC08 Dectape units 0–7. The other version copies files among disk units 0–3 and TD8E Dectape unit 0, and calls the program MAGCPY which copies files to and from 9 track industry standard magnetic tape. If a transfer to or from industry standard magnetic tape is desired, a M response is made to these prompts. The industry standard magnetic tape must be unit 0. To copy files to an industry standard magnetic tape, a write enable ring must be installed on the tape reel. This function is accomplished with a Dectape by pressing the write enable switch.

A system disk or Dectape may be duplicated (copied in its entirety) with the program CPYDSK if two disk drives or two Dectape drives are present, which implies a TC08 Dectape. The dialogue for this program is as follows:

```
SELECT MODE: CPYDSK
FROM DISK? Y
UNIT NO.: 0
ONTO DISK? Y
UNIT NO.: 1
```

```
SELECT MODE: CPYDSK
FROM DISK? N
TAPE UNIT NO.: 1
ONTO DISK? N
ONTO TAPE UNIT NO.: 0
```

Industry standard magnetic tape to industry standard magnetic tape duplications are not supported.

A very versatile and useful program called SAVE is available to transfer parts of data files from one system disk or Dectape to another, or to segment data files on a single disk. The original intent of this program appears to have been to give the operator the capability of saving a collection of mass spectra of various compounds using different spectra selected from different data files. However, the program SAVE has broader applicability and may be used to reduce the size of data files by saving, under a new file name, contiguous groups of spectra that correspond to the parts of a chromatogram that contain the peaks. The original data file is then deleted and many useless spectra of leading, interspersed, or trailing baseline are purged from the storage device. It is very important to recognize that the SAVE program does not maintain the original spectrum index numbers, but renumbers the spectra in the saved file. Therefore all sense of retention time is lost, and if this information is desired, a TICP of the original data file should be plotted before executing the SAVE program. In the sample dialogue shown below, the SAVE program is used to segment a data file containing 372 spectra but contains all the relevant information except retention time. All of the conventional output programs described in chapter 4 may be used with the reduced and saved data file to plot TICP's, EICP's and mass spectra.

```
SELECT MODE:  SAVE
FILE NAME:  RW147B
RENAME:  Y
NEW FILE NAME:  SRW147
FROM DISK?  Y
UNIT NO.: 1
ONTO DISK?  Y
UNIT NO.:  0
SPECTRUM NUMBERS:
170-200;230-290;300
```

In this example the original data file was on disk unit 1 and the new data file is saved on disk unit 0. It is not necessary to transfer a file to another unit with the SAVE program. If the original and saved data files have different names, they may be located on the same device. A negative response to the ONTO DISK? prompt causes the program to enter the Dectape dialogue, and the new data file may be saved on Dectape. The SAVE program does not support an industry standard magnetic tape. The user should be aware that there are two version of SAVE; one handles TC08 Dectapes 0–7 and disks 0–3; the other handles TD8E unit 0 only and disks 0–3.

In the same dialogue a single spectrum No. 300, was saved illustrating this capability. It should be noted that there is no background subtract capability in the SAVE program and it is the background subtracted spectrum that should be saved in any collection of individual mass spectra. Therefore the feature of the plotter output routine (Chapter 4) that allows the operator to save a subtracted file should be used first and the background subtracted spectrum saved with the SAVE program.

FILE DELETE AND SYSTEM TAPE DUMP PROGRAMS

The program DELE is a housekeeping program which removes programs or data files from the disk or Dectape directory. It is essential for a GC/MS operator to maintain an orderly storage system and delete all unneeded files. The dialogue for the program is:

```
SELECT MODE:  DELE
FILE NAME:  FILE TO BE DELETED
TAPE?:  Y (if the file is on tape)
        N (if the file is on disk)
```

This program operates most efficiently if files at the end of the list are deleted first, followed by older files. The user should be aware that there are two versions of DELE; one handles TC08 Dectape unit 1 and disk 0; the other handles TD8E unit 0 and disk 0.

An alternative program, BCDELE, is more efficient in deleting files, but requires the presence of an extended arithmetic element (EAE) and 8K core in the PDP-8. More information on the EAE is included in section 4.2 on the MSSOUT software. BCDELE will remove from disk unit 0 only contiguous groups of files as in the following example:

```
MOUTP4 7431
SAMP1  7457
SAMP2  7465
SAMP3  7473
SAMP4  7501
SAMP5  7507
SAMP6  7515
SAMP7  7523
SAMP8  7531
SAMP9  7537
SAMP10 7545

SELECT MODE:  BCDELE
*SAMP1:SAMP2
*SAMP5:SAMP6
*SAMP8:SAMP9
*  press return
```

```
MOUTP4 7431
SAMP3  7457
SAMP4  7465
SAMP7  7473
SAMP10 7501
```

System dump programs cause the entire contents of the system disk to be copied to Dectape or industry standard magnetic tape in a form that is suitable for their recopying by another program onto a formatted, but otherwise empty disk cartridge. The value of this program is in creating a number of new disk cartridges containing identical system software from a single dump tape. Sample dialogue for the SYSDMP program, which dumps the system to Dectape, is:

```
SELECT MODE:  SYSDMP

7 TAPES WILL BE NEEDED
MOUNT REEL 1 ON UNIT 1
```

The corresponding program for industry standard magnetic tape is called MAGDMP. The method of restoring dump tapes to system cartridges is discussed under the OS-8 operating system. Again the user should be aware that there are two versions of SYSDMP, one for TC08 Dectape units 0–7 and one for the TD8E Dectape unit 0.

PROGRAM TO CHANGE MACHINE LANGUAGE INSTRUCTIONS

The program PATCH is an extremely dangerous one because it allows the user to change machine language instructions in programs as they are stored in disk files. The program is important because certain changes are sometimes essential. An example is the change of status bit 10 in location 2060 of INIT$$ to permit use of M1000 to calibrate with tris(perfluoroheptyl)triazine. The procedure for this change is as follows:

SELECT MODE: <u>PATCH</u>
FILE: <u>INIT$$</u>
(computer halts)

Examine location 2060 and modify to assure that bit 10=1; for example, load address 2060, examine contents (this process bumps the address to 2061); load address 2060, set all switch registers to 0 or 1, press deposit key;

set switch register to 0200, press load address, clear, and continue. The system will boot.
Bits in location 2060 have the following significance:

bit	0=no 1=yes
0	Second TC08 Dectape for data acquisition
1	TC08 Dectape system
2	Dual drive TC08 Dectape
3	Disk system
4	Dual disk system
5	IFSS interface
6	Plotter
7	TD8E Dectape or 9 track tape or cassette
8	CRT
9	Disk monitor in core
10	1000 amu file handling
11	One millisecond timing board (fast scan option)

PROGRAM TO START THE OS-8 SYSTEM

The OS-8 system may be started by inserting an OS-8 disk cartridge, and then entering into the switch register a bootstrap loader. An alternative method is to start a system program, then insert an OS-8 disk cartridge. The system program is called OSDGO and sample dialogue is as follows:

```
SELECT MODE:  OSDGO

PLACE OS/8 CARTRIDGE IN DRIVE;
WAIT 'TIL READY; THEN
PRESS CONTINUE
```

7.2 THE OS-8 OPERATING SYSTEM

If a real time system disk is not up and running, it is simpler and quicker to start OS-8 by entering the OS-8 bootstrap program into the switch register and running it in the usual manner. See section 2.3 for instructions on loading a program into the switch register. If a system disk is up and running, it is simpler to start OS-8 with the program described in the previous section. The OS-8 bootstrap is as follows:

Address	Instruction
0000	6502
0001	0000
0002	6517
0003	6512
0004	6514
0005	5005

If this program runs successfully, a period will be printed on the console output terminal. If the program fails to run properly the first time, it must be reloaded and restarted. This short bootstrap is of the type that is destroyed as the OS-8 system loads, but a new bootstrap is created at address 7600. Therefore after a successful start if it is necessary to halt the computer, OS-8 may be restarted at address 7600.

An extensive discussion of the OS-8 operating system and many of the programs found on the OS-8 system is well beyond the scope of this chapter or this manual. Users who desire additional information should consult the OS-8 handbook (Digital Equipment Corporation part number DEC-S8-OSHBA-A-D, about $5.00). The only programs discussed in this chapter are those relevant to GC/MS system operations and maintenance.

TAPE FORMAT PROGRAMS

Dectape must be formatted using a program analogous to the disk cartridge format program. Two programs are available named DTFRMT and TDFRMT and these apply to Dectapes using the TC08 and TD8E controllers respectively. Mount the Dectape to be formatted on the drive and select a unit number with the thumbwheel switch. Turn each tape twelve turns onto the take-up reel, place the WRITE LOCK/WRITE ENABLE switch in the write enable position, and the LOCAL/REMOTE switch in the remote position. The NORMAL/WRTM switch is on the front of the TC08 tape controller behind the facing panel and this must be placed in the WRTM position. With the TD8E controller the corresponding switch is located on the upper right corner of the M868 board on the PDP-8 bus. It is necessary to slide the PDP-8 forward and remove the lid to see the switch. Its normal position is off and the format position is labeled WTM. Call the format programs under OS-8 as follows:

```
.R DTFRMT (TC08 controller)

DTA? 1
DIRECT? MARK
```

```
0201 WORDS, 2702 BLOCKS. OK?
(YES OR NO)
YES
SET SWITCH TO NORMAL

.R TDFRMT (TD8E controller)

UNIT? 0
FORMAT? MARK
0201 WORDS, 2702 BLOCKS. OK?
(YES OR NO)
YES
SET SWITCH TO WTM
(after one pass down the tape)
SET SWITCH TO NORMAL
```

The response to the first prompt is the Dectape unit number selected. On a dual drive system two unit numbers may be entered to format two dectapes. After one pass of the tape the message SET SWITCH TO NORMAL is printed. Set the switch to NORMAL (or OFF) and press the return. At the conclusion of the formatting, the program returns to the second prompt and additional Dectapes may be formatted. After all the Dectapes are formatted, halt the program and restart OS8 at address 7600.

PROGRAM TO LIST FILES ON THE OS-8 SYSTEM

There are several methods of listing the names of the files on the OS-8 system. These are described in detail in the OS-8 handbook, and only one method is described briefly in this section. One point to remember is that the OS-8 system disk is divided into two parts, a system area and a default storage area. Files in both areas may be listed with the program PIP as follows.

```
.R PIP
*SYS:/E
*/E
```

Entry of the command /E without designating the SYS: area causes the program to list the files in the default storage area.

PROGRAMS TO BUILD REAL TIME SYSTEMS FROM TAPE DUMPS

A real time system tape dump contains all the files necessary to regenerate a new disk. These dumps are prepared using the programs

SYSDMP or MAGDMP as described in section 7.1. A single dump tape may be used to generate a number of system disks. Three programs are available which are named TC8RES (TC08 Dectape controller), TD8RES (TD8E Dectape controller), and MAGRES industry standard magnetic tape).

The two Dectape programs operate very much the same. In each case the first reel of the dump tape (there may be one or more reels depending on the number of data files included in the dump) should be mounted before calling the program. If a TC08 Dectape controller is present, mount the tape on unit 1; if a TD8E controller is present, select unit 0. The Dectape drive should be in the write lock position to protect the dump tape in case of an error. The sample dialogue is as follows:

```
.R TC8RES          (TC08 controller)
.R TD8RES          (TD8E controller)

MOUNT REEL 1 ON UNIT 1
(TC08 controller)
MOUNT REEL 1 ON UNIT 0
(TD8E controller)
```

At this point remove the OS-8 cartridge and install a formatted disk cartridge. This cartridge may contain files and programs, but they will all be replaced by the real time system files. Press the return when the disk is up to speed. After the restoration is complete, the operator may start the new system by responding Y to the START SYSTEM prompt. A negative response causes the computer to halt. Another cartridge may be restored without reinserting the OS-8 system. Place another formatted cartridge in the drive and restart the restore program at address 0200.

The MAGRES program is used very much the same. The write ring should be removed from the dump tape and the disk controller format switch on the rear of the disk controller should be placed in the format position.

LOADING NEW PROGRAMS ON OS-8 OR THE REAL TIME SYSTEM

Programs may be transferred from the OS-8 system to the real time system and run under this system. Similarly programs may be loaded onto the OS-8 system from OS-8 Dectapes or paper tape. Finally new programs may be written in PDP-8 assembly language with the use of the OS-8 operating system, and transferred as above. All of these transfers are possible using a variety of OS-8 options, and is well beyond the scope of this manual to describe all of the possibilities. The user should consult the OS-8 handbook for detailed documentation. This section does present instructions for a few simple transfers that are especially valuable.

Inserting a Paper Tape Handler in OS-8. Some OS-8 system disks are not set up to handle paper tape input from a teletypewriter. This insertion is required just once and is accomplished by entering the following commands to the OS-8 monitor:

```
.RUN SYS BUILD
$INSERT KS33,PTR
$ BOOT
SYS BUILT
.SAVE SYS BUILD
```

Loading a Paper Tape File on OS-8. Paper tape files come in two basic varieties: binary and ASCII formats. Binary tapes are usually labeled FILENAME.BN and ASCII tapes are labeled FILENAME.PA. One may load a program from a binary file onto the OS-8 system and run it with the OS-8 R command by using the following procedure:

```
.R ABSLDR
*PTR:
(insert paper tape into the reader in the region where only the
parity bit is punched; start the reader)
```

The program will print an asterisk after reading the tape. Stop the paper tape reader and enter a control/C. After the period is printed, enter the following command:

```
.SAVE SYS: FILENAME
```

Either ASCII or binary paper tape files may be transferred to the OS-8 system area with the program PIP. If the file is in ASCII format, use the /A option; if binary use the /B option.

```
.R PIP
*SYS: FILENAME.PA < PTR:/A
(insert paper tape into the reader
In the region where only the parity
bit is punched; start the reader)
```

The program will print an asterisk after reading the tape. Stop the paper tape reader and enter a Control/C.

Loading a Dectape File on OS-8. An OS-8 Dectape file may be transferred to the OS-8 disk by several programs. The program FOTP is convenient since it automatically generates the same filename on the disk as was present on the Dectape, and several files may be transferred with one command:

```
.R FOTP
*SYS: < DTA1:Filename.BN
```

This command would cause a binary file on Dectape unit 1 to be transferred to the OS-8 disk. Under the OS-8 system Dectape units 0 and 1 are usually reserved for TD8E controllers and Dectape units 2 and 3 are usually reserved for TC08 controllers. More information on this program is found in the OS-8 handbook.

LOADING A BINARY PROGRAM ON THE REAL TIME SYSTEM DISK

A binary file program on the OS-8 system may be transferred to the real time system disk with the following procedure:

```
.R ABSLDR
*SYS:    FILENAME,LOD150/G
```

This causes the file to be loaded on the real time system to be placed in core memory; a second binary file called LOD150 is also loaded and started by the /G option. LOD150 is a program that has several other versions named DLOADO and LOADIT. LOD150 produces the dialogue:

```
FILE NAME:  FILENAME
P.A.:  enter page addresses
P.A.:  e.g., 0200-7577
P.A.:  0020-0157
ONTO UNIT NO.:  0 (1 if Dectape)
ONTO TAPE?:  N    (Y if Dectape)
```

Before pressing the return remove the OS-8 disk and insert the real time system disk. Press return when ready and the message FILE HAS BEEN LOADED will be printed at the completion of the transfer.

PROGRAM TO DUPLICATE OS-8

The program CPYDSK, described in 7.1, will duplicate the OS-8 system. On OS-8 disks it is listed as DSKCPY, but the dialogue is the same.

PROGRAM TO START THE REAL TIME SYSTEM

The OS-8 program MSDGO will start a GC/MS system disk as follows:

```
.R MSDGO
PUT S/150 CARTRIDGE IN DRIVE;
WAIT 'TIL READY; THEN
PRESS CONTINUE
```

CHAPTER 8
PREVENTIVE MAINTENANCE

The purpose of this chapter is to gather together in one place preventive maintenance information for the complete computerized GC/MS system. As in several other chapters, some detailed information is specifically oriented to the Finnigan quadrupole mass spectrometer with a data system that uses a Digital Equipment Corporation model PDP-8 computer. However a significant amount of information is of a general nature and is applicable to any computerized GC/MS system. The sections in this chapter do not repeat detailed instructions that are adequately covered in manufacturers' manuals. The user is referred to these instructions in the appropriate manual.

Preventive maintenance is perhaps more important with a mass spectrometer than any other instrument system. This is because the mass spectrometer vacuum system is turned on 24 hours per day, 7 days per week. The failure of a relatively simple component at an inappropriate time can result in the total destruction of an expensive and vital part of the system and therefore make the total system useless.

8.1 MASS SPECTROMETER VACUUM SYSTEM

MECHANICAL PUMPS

The oil in mechanical pumps must be changed occasionally. The frequency of oil changes is determined by the number of samples introduced that contain high concentrations of materials. Heavy use of a direct inlet system invariably leads to rapid contamination of the pump oil. In general for a heavily used system, the mechanical pump oil should be changed at least once a year. If the system is used mainly for GC/MS work with relatively clean samples, a longer period may be acceptable. Similarly, heavy use of a direct inlet system may require an oil change every six months. The direct drive pumps may require attention every six months regardless of the sample load. Edwards No. 16 oil and Welch DUO-SEAL oil are acceptable as pump oils. Direct drive mechanical vacuum pump fluid should be used in the Alcatel pumps.

Mechanical pump oils should be purged occasionally for a few seconds to vent impurities. This is accomplished by opening the oil reservoir cap and allowing the pump to gulp air for a few seconds. A mechanical pump pressure of 0.1 Torr or higher after purging is an indication that an oil change is required. A clean mechanical pump operates at 0.01 - 0.05 Torr. Unreliable pressure measurements can result from dirty contacts on the pressure gauge switch. Spray contact cleaner should be used as required to insure accurate pressure measurements.

The mechanical pump oil level gauges should be inspected periodically and cleaned if necessary. Soak the tubes in acetone for 30 minutes, rinse with methanol, and dry with a heat gun. Some pumps have small windows instead of tubes for observing oil levels. These cannot be cleaned. Tygon vacuum lines may be cleaned with methanol but not with acetone which dissolves Tygon. This should be done whenever the oil is changed and the lines should be inspected for soft spots or wear, particularly at the connection points. The metal tube connecting the pump to the separator should be cleaned at the same time since oil can collect in the horizontal part of the fitting and contribute to a high background. Metal fittings may be washed with acetone and rinsed with methanol.

Mechanical pump belts should be checked for correct tension and wear at least every six months or any time the system is shut down. A correctly adjusted belt can be twisted 90 degrees. The rough pump belts often give a warning signal when they start to fray or develop cracks prior to breaking; they usually develop a noise and the operator should be aware of any change in the sound of the pumps. If the pump guards are removed and left off, it is much more conducive to regular checks on the condition of the belts.

Special attention should be given to oil leaks if they develop. On direct drive pumps the motor to pump seals may be replaced if necessary.

DIFFUSION PUMPS

Oil changes are required for diffusion pumps and the considerations for the frequency of changes are the same as discussed under mechanical pumps. In general, diffusion pump oils should remain cleaner than mechanical pump oils because the mechanical pumps are at the end of the vacuum system. Diffusion pump oil can be used for up to three years if the pressure remains good and the background low. Use Santovac No. 5 oil (a polyphenyl ether made by Monsanto) in the main analyzer diffusion pumps. This oil has a very low vapor pressure and its background contribution is negligible. A silicone oil such as Dow-Corning 704 or 705 is used in the smaller diffusion pump attached to the batch and direct inlet systems. This oil has a greater resistance to oxidation and is better suited to the inlet systems where oxygen pressures may be high. Whenever a diffusion pump oil is changed, the old O-rings should be replaced. Tygon tubes should be inspected and cleaned as described under mechanical pumps.

The cooling water exits of the diffusion pumps should be checked daily for proper flow. Occasionally solid particles from corrosion plug the exit fittings from a pump and cause it to overheat. This is more likely to happen in an older system. If flow is bypassing one pump, the cooling coils will be hot to the touch. The cooling water lines should be cleaned whenever the pump oil is changed. A complexing agent such as EDTA (Versene) or other commercial cleaning agents work well for this. If cooling water is 50 ppm calcium hardness or higher, lines will plug within two years if not cleaned. If the cooling water has a high sediment content, a reliable sediment filter is recommended for the cooling water line. This filter should be maintained regularly to prevent reduced flow rates.

ION GAUGE, MANIFOLD, AND INLET SYSTEMS

The Bayard-Alpert ion gauge should be outgassed daily for approximately five minutes. The ion gauge may be wrapped in aluminum foil to increase the temperature of the tube. This helps keep the gauge clean and expedites outgassing. Some Finnigan models are fitted with an aluminum protective cover for the ion gauge. This also serves as a reflector to increase the tube temperature. If this shield is not provided, it is a good idea to build a plexiglass cover for the ion gauge. A dropped tool and a suddenly shattered ion gauge tube is a catastrophic event.

The manifold may be baked out occasionally at 200-250C if background is a problem. The magnet should be removed and the ionizer and electron multiplier potentials should be turned off. In the event of extreme contamination, the manifold can be cleaned with acetone or methanol.

If the direct inlet system is used frequently, pump oil can backstream and settle in the solid probe area and contribute to background. The direct inlet manifold may be cleaned with a cotton swab dipped in acetone or methanol. However care must be exercised in baking the system since the direct inlet valves contain Viton O-rings. These O-ring seals should be replaced at regular intervals if the direct inlet system is used frequently.

The metal disk filter between the GC column and the separator must be cleaned if the gas chromatograph is operated at high temperatures. The frit is cleaned with methylene chloride in an ultrasonic bath.

The manifold, batch inlet, transfer line, and separator oven temperature monitoring systems must be verified occasionally with a reliable thermocouple and a good external potentiometer. The contacts on the temperature switch should be cleaned with an abrasive paper and spray contact cleaner to insure accurate temperature readings.

The batch inlet system septum should be changed each time the system is brought to atmospheric pressure.

8.2 THE MASS SPECTROMETER ELECTRONICS

COOLING FANS AND DUST FILTERS

With modern solid state electronics the single most important preventive maintenance factor is the flow of cool air across and around electronic components. Clean dust filters and working fans provide these flows and it is absolutely essential to maintain them properly. The proper operation of cooling fans is extremely easy to check and this should be done once a week. The time interval for cleaning cooling fan filters varies widely depending on location and the air quality in the mass spectrometer laboratory. Frequent checks of air flows and the cleaning of filters quarterly is essential.

The Finnigan model 1015 has cooling fans with filters on the oscilloscope, RF/DC generator, and power supply modules. A fan without a filter is contained in the oscillographic recorder. Other models have a similar placement of fans and filters.

VACUUM TUBES

The older Finnigan model 1015 systems have in excess of 25 vacuum tubes and the 3000 series instruments have four vacuum tubes in the RF/DC generator (one 6AL5, one 6EB8, and two 8458). There is wide disagreement about changing vacuum tubes on a regular basis in any preventive maintenance program. One extreme philosophy is to never change a tube until unstable operations or failures occur. At that time tubes are prime trouble shooting targets. It is more prudent to regularly change tubes and the schedule suggested in this manual is intended as a guideline for the 1015 system.

RF/DC Generator. Replace 5726 and 807 tubes annually. The 5726 is a 6AL5 replacement.

Oscilloscope. Replace the 6BQ6 tube annually.

Electron Multiplier Power Supply (Keithley). Replace the 8068 and OA2 tubes annually.

Low Voltage Power Supply (Dressen-Barnes). Replace the 6AV6, OA2, 12AX7, 655OA, and 5R4GYB tubes annually. Check 655OA and 5R4GYB tubes semi-annually.

Low Voltage Power Supply (Finnigan). Replace the 6LQ6, OA2, and 12AX7 tubes annually.
Oscillographic Recorder (Century). Clean the optical surfaces annually. These are cleaned in place with a cotton swab and alcohol; wipe clean with a tissue. Check the condition of the drive belt and the galvos annually.

8.3 THE GAS CHROMATOGRAPH

The GC septum should be changed weekly. The oven and injection block temperature monitoring systems must be verified occasionally with a reliable thermocouple and a good external potentiometer. The dirt filter should be cleaned regularly. The utilization of drying agents, filter traps, and an oxygen scrubber is recommended for the carrier gas line, and these should be regenerated or replaced as necessary.

8.4 THE DATA SYSTEM

SOFTWARE

Computer programs are called software because they can be changed relatively easily in contrast to the resistors, transistors, and integrated circuits of the hardware. This capability of relative ease of change of software can provide enormous flexibility and the possibility of continuous system updating and improvement. It also can create a chaos of magnetic tape and disk cartridges whose contents are all different and unknown to the user.

The very first consideration of software preventive maintenance is an organized, systematic method of dating and labeling all paper tapes, magnetic tapes, and disk cartridges. Similarly all documentation, which may be received as loose update sheets, should be dated and stored in a single file for ease of reference. All operating system software should be duplicated before any attempt is made to run the programs. Instructions for duplicating system disk cartridges, OS-8 disk cartridges, and Dectapes are given in chapter 7. It is a sound policy to make several copies of important disks and tapes and place one copy of each in a secure cabinet. These copies should be reserved for ultimate back-up that is never used for anything except making copies.

COOLING FANS AND DUST FILTERS

The considerations discussed in section 8.2 are equally applicable to the data system. The PDP-8 data system contains numerous fans and filters.

All floor standing cabinets contain a double fan and a large filter at the rear cabinet base. The disk controller and disk controller power supply modules also have fans and filters. The PDP-8 has an internal fan and filter. The Dectape drive and its power supply have cooling fans. The Houston plotter has a cooling fan and the Tektronix 4610 hard copy unit has a dust filter. The System Industries digital interface has a cooling fan and dust filter. The RIB interface has no fan or filter.

LUBRICATION OF MECHANICAL PARTS

Because of the mechanical complexity of the teletype, an annual service call or service contract is recommended. If service is not readily available, the motor should be oiled, the motor gears greased, the belt checked, and dashpot and dashpot cylinder cleaned and reoiled at least annually. It is recommended that the teletype be turned off at all times when not in actual use.

The plotter's sliding metal surfaces should be cleaned quarterly with isopropyl alcohol. Do not oil or grease any moving parts or sliding surfaces; all moving parts are permanently lubricated or made of low friction combinations.

The slide mechanisms of electronic modules may be lubricated with machine oil or graphite as needed.

MAGNETIC STORAGE DEVICES

The read/write heads of the magnetic tape and disk drive units must be kept clean and free of dirt and iron oxide. The tape heads are readily cleaned with isopropyl alcohol and a soft tissue monthly. The disk drive heads should be cleaned semi-annually with isopropyl alcohol and a lint free wiper. Further information is contained in the disk drive manual.

DIAGNOSTIC COMPUTER PROGRAMS

Certain diagnostic tests are utilized for troubleshooting, and several of these are recommended for checking the integrity of the system on a routine basis. These tests are explained in detail in chapter 9.

CHAPTER 9
TROUBLE SHOOTING

Trouble shooting is often the most difficult and frustrating part of operating a computerized GC/MS system. The failure of an extremely simple component or electrical connection may be the cause of a complete failure of the system. However, because of the complexity of the total system, it may be extremely difficult to isolate the simple problem.

The purpose of this chapter is to develop a systematic method of trouble shooting a computerized GC/MS system. As in other chapters, some sections are oriented to Finnigan model 1015 and 3000 series quadrupole mass spectrometers with data systems that employ Digital Equipment Corporation model PDP-8 computers. However, the overall systematic approach is applicable to any GC/MS system.

The chapter is divided into five sections. The order of the sections is the same as the probable order of discovery of problems by an operator following the recommended operational procedures in chapter 2. Within each section the problem solving approach is the same. The first things to look for are simple, easy to fix problems such as loose or broken cables, blown fuses, dirty PC board connections, inoperable cooling fans, dirty dust filters, water or oil leaks, air leaks, and uncontrolled heaters. If this fails to locate the difficulty, it is necessary to isolate the problem as much as possible to a specific module or modules. Once this is accomplished, manufactures' service manuals should be consulted to determine specific remedies. It is beyond the scope of the manual to catalog a large number of specific failures and remedies. There are simply too many variations in hardware even within one manufacturer's product line. If the manufacture's service manual does not provide a solution to the problem, it is then appropriate to call a service engineer for the specific module that is suspect.

9.1 MASS SPECTROMETER PROBLEMS

VACUUM SYSTEM

There are several warning signals that should be checked each day on entering the mass spectrometer laboratory. A diffusion pump over heat light will come on if the cooling water flow was too low or the water insufficiently chilled to maintain the pumps at a proper temperature. A Fenwall switch senses the temperature at each diffusion pump and automatically shuts off the high vacuum system in the event of overheat. If the overheat is limited to just one diffusion pump, the entire system is shut down. The remedy for this problem is clearly to adjust the cooling water flow to a correct level and to assure that the outlet stream is cool. The Fenwall switch may require resetting before the high vacuum system is reactivated. See chapter 2 for the pump down procedure.

Another warning signal is the illumination of the pressure failure light or an indication of an abnormally high pressure on the vacuum gauge. With the GC column disconnected or the carrier gas diverted from the mass spectrometer, a pressure of 5×10^{-7} to 10^{-8} Torr should be maintained. Under these conditions air leaks are insignificant and the presence of ions of masses 18,28,32, and 40 are of no consequence. If the pressure becomes sufficiently high, about 10^{-4} Torr, the high vacuum system will be shut down. The most common causes of vacuum system failures are a broken belt on a mechanical pump, a break in a Tygon or glass vacuum line, or a leaking O-ring or Teflon ferrule. Teflon ferrules flow when hot and the metal fittings need to be tightened occasionally. Electrical problems may also cause high vacuum system failures. A momentary power failure will shut down the high vacuum system but the mechanical pumps will restart when the power returns. A blown fuse or a faulty component on the ion gauge controller board may also cause a pressure failure.

For abnormally high pressures that do not shut down the vacuum system, the first remedy is to tighten metal fittings containing Teflon ferrules about one-half turn. If this fails to eliminate the leak, inert gas such as argon or helium should be sprayed around fittings, gaskets, tube connections, and O-rings. The observation of a sudden increase in pressure is an indication of a leak. If this procedure fails to detect the leak, turn the ionizer and electron multiplier on and set the first and last mass controls to display the prominent ions from the inert gas. Continue spraying the gas and look for a sudden increase in ion abundance at the characteristic masses. It may be necessary to change a gasket or leaking O-ring.

ION GAUGE

The glow of the Bayard-Alpert tube is an indication of the state of the vacuum system. If the tube glows extremely brightly, flickers intermittently, or flickers rapidly when the solvent from a GC injection reaches the manifold, trouble is indicated. Similarly, if the high vacuum gauge seems insensitive to changes in pressure as during solvent elution or the admitting of PFTBA, the ion gauge should be serviced.

Intermittent flickering may be caused by contamination from solvents or pump oil. Degassing overnight may help as will the application of external heat with a heat gun. See also chapter 8 for preventive maintenance measures. For other problems, the ion gauge controller board should be examined for bad components. If this fails to solve the problem, replace the vacuum tube.

HEATERS

Manifold, separator, transfer line, or batch inlet heaters may lose control and overheat. If this occurs, the respective triac controllers are suspect and should be tested and replaced. If higher than normal potentiometer settings are required to maintain temperatures, one-half of a clamshell heater may be burned out. Also, bad thermocouples and dirty meter switch contacts cause inaccurate temperature readings.

9.2 DATA SYSTEM PROBLEMS

Many data system problems are caused by simple failures that are readily corrected. Power cables and signal transmission cables may work loose and all such connections should be checked at the first sign of trouble. Similarly fuses, cooling fans, and filters should be checked at once. Some modules have normal/test switches and if a switch was inadvertantly left in a test position, the module may not funtion normally.

Magnetic read/write heads should be cleaned as described in the preventive maintenance chapter. If software on a disk or tape was altered by some malfunction or accident, programs may suddenly fail. An alternative disk or tape of known good quality should be tried. If these measures fail, it is advisable to check all PC boards for secure, tight, and clean connections. It is worth lifting each PC board in a unit from its connector, removing dirt corrosion with an ordinary pencil eraser, and replacing each board.

Electronic component and some mechanical failures are best found with the use of a diagnostic computer program. The programs in the following sections are for the PDP-8 data system, but comparable programs should be made available by each data system manufacturer. After isolation of the particular problem with the diagnostic program, it is appropriate to call a service engineer for the faulty module.

COMPUTER DIAGNOSTIC PROGRAMS

A large number of computer diagnostic programs are available for the PDP-8 from the Digital Equipment Corporation. Each of these programs tests the operation of one or several computer functions. Digital Equipment Corporation maintenance and diagnostic programs are recognized by the characteristic name, MAINDEC, followed by a series of digits and letters. Each progam is supported by a document that describes the use of the program and the error messages that are printed. These documents are required to gain full advantage of the diagnostic programs.

The most important decisions that must be made when trouble shooting the data system are which diagnostic program to run and when to call the service engineer. In general it is possible to save considerable time and expense by running diagnostics to isolate the problem before calling a service engineer. Reporting the results of the diagnostics at the time the service call is made sometimes allows a quick solution to the problem by replacement of a specific printed circuit board. In a few cases the choice of a diagnostic program is quite clear. For example, if a problem repeatedly occurs when using a magnetic tape drive and only when using the tape, find a diagnostic for the magnetic tape system and run it. Dectape diagnostics are discussed in the next section of this manual. If all the more readily recognized device problems are eliminated, the magnetic disk system is the next peripheral to check since it contains mechanical funtions that are more susceptible to failure than the purely electronic functions of the computer itself. The disk diagnostic program is described in a subsequent section and it is not a standard Digital Equipment Corporation program. The PDP-8 disk system employs a Diablo Corporation series 30 moving head disk drive and a controller manufactured by System Industries. The diagnostic program was written by System Industries and it has its own detailed documentation.

If it is determined that the disk system is functioning properly, it is appropriate to run computer diagnostics. Table 9.1 contains a list of OS-8 program names for a selection of computer diagnostic programs, brief operation instructions, and the satisfactory test indications. This list is a small subset of the total number of available computer diagnostics, but it includes some of the most general tests that are recommended for reasonably comprehensive testing of most basic computer functions. It is strongly recommended that the GC/MS operator run several or more of these programs at regular intervals in order to become familiar with their correct operation. With systems that are not used often, these programs should be run regularly to exercise the system and keep it in good operating condition. With these diagnostic programs stored on the OS-8 operating system, it is easily possible to test the computer's function in a few hours. If the complete program documentation is not available, it is possible to gain some valuable information with the general directions that are given in Table 9.1. To load the programs from the OS-8 system, enter R, then a space, and the name of the diagnostic program. The programs should be run in the sequence shown in Table 9.1 since some of the later tests assume correct functioning of earlier tests. All programs have the standard starting address 0200 if it is necessary to restart them. The programs are stopped by pressing the HALT switch. The XADDR, XCHKBD, and EAEXME programs require at least 8K of memory, but there are corresponding programs for 4K systems. Error messages are printed in the event errors are found, and the complete program documentation is required to interpret the error numbers. The error number explanations may have little meaning to most operators, but should be available before calling a service engineer. After halting the program, OS-8 may be restarted at address 7600. However the two memory diagnostics write over all contents of memory, and the OS-8 bootstrap must be reloaded to restart OS-8 in these cases.

For problems that are intermittent, it is recommended that one of the memory or general instruction test programs be permitted to run overnight. Raising the temperature of the room during testing is a method of causing marginal components to fail completely.

A complete list of the MAINDEC diagnostic programs is found under acceptance tests in the Digital Equipment Corporation PDP8/E, PDP8/F and PDP8/M Processor Maintenance Manual – Volume 1. Additional tests for modules not included in Volume 1 are found in Volume 2, Internal Bus Options or in Volume 3, External Bus Options. In Volumes 2 and 3 the available MAINDEC programs are usually listed under Software, Acceptance Tests, Maintenance, or Trouble Shooting. Lists of maintenance and diagnostic programs are also found in other Digital Equipment Corporation publications. For example a list is found in appendix A of the 1971 edition of the PDP8/E Small Computer Handbook.

DECTAPE DIAGNOSTIC PROGRAMS

The Dectape diagnostics are described separately because problems may be readily identified with this unit when failures occur only during Dectape operations. Also included is a relatively simple tape drive brake ajustment that is a solution to several problems.

There is one aspect of Dectapes that causes confusion. All Dectapes are not alike although they may appear physically identical and data files and programs may be transferable between them. The most common Dectape drive is the TU56 in either a single or double tape drive configuration. The confusion arises because this drive is used with the relatively slow TD8E interface board and the relatively fast TC08 Dectape controller. The TD8E and the TC08 are not compatible and Dectape access programs written for one will not work with the other. Each controller has a different format

Table 9.1 Computer Diagnostic Programs for the PDP-8/E or M

OS-8 Program Name	Function	Operation	Satisfactory Test Indication
INST1	CPU instruction tests	halts on loading; enter 7777 in switch register and press clear and continue	bell every few seconds
INST2	CPU instruction tests	starts on loading	bell every few seconds
XADDR	extended memory address tests	halts on loading; enter starting address 0200 in switch register, press load address, clear, and continue; halts and prints message; enter in the switch register 0001 for 8K, 0002 for 12K, etc. and press continue	prints the numeral 5 every five minutes
XCHKBD	extended core memory tests	halts on loading; enter in the switch register 0200, press load address, clear and continue; halts and prints message; enter in the switch register 0001 for 8K, 0002 for 12K, etc. and press continue	prints the numeral 5 every five minutes

ADDER	CPU adder circuit tests	starts on loading; before calling program enter in the switch register 0210 for 8K, 0220 for 12K, etc and then call program	prints the characters SIMAD, SIMROT, FCT, and RANDOM after 35 minutes
EAEST1	extended arithmetic element instruction tests	halts on loading; enter 0200 in switch register, press load address; enter 1000 in switch register; press clear, and continue	prints the characters KE81 every 6 minutes
EAEST2	extended arithmetic element instruction tests	halts on loading; enter 0200 in switch register, press load address, clear and continue; halts; enter 0400 in switch register and press continue	rings bell every minute
EAEXME	extended arithmetic element – memory tests	halts on loading; enter 0200 in switch register and press load address; enter 4000 in switch register for 4K, 4001 for 8K, 4002 for 12K, etc; press clear and continue; after message enter 0 for 4K, 1 for 8K, 2 for 12K, on the terminal keyboard and press return	prints the characters KE8 EME every 22 seconds

program and different diagnostic programs but data files and non-Dectape programs are fully compatible with each system. To add to the confusion, older documentation often refers to the older TU55 tape drive and TC01 Controller which are very similar to the TU56 and the TC08.

The diagnostic programs for Dectape units are listed in the appropriate maintenance manual, e.g., the TC08 Dectape Controller maintenance manual under Software. If a Dectape is suspected of causing transmission errors or exhibits sluggish performance, an exerciser program should be run. A random exerciser program called TC08RX is available for the TC08 Dectape controller. A program called TD8DIA is available for TD8E Dectapes. The use of the TD8E test program requires the complete program documentation and this program is not discussed here. The following information refers to the TC08 test program, but TD8E users should take note of the TU56 brake drive adjustments and other comments in the last two paragraphs of this section.

After the program is loaded into memory, the computer will halt. Mount a formatted Dectape on the drive or drives to be tested. More than one drive may be tested at the same time if desired. The tape or tapes must not contain any data or programs that are needed, as the diagnostic destroys all information on the tape. Set the desired drive number using the thumbwheel switch on the Dectape drive and place the unit in write enable and remote. Load the starting address of the program, 0200, into the switch register and press LOAD ADDRESS. Set the switch register to select the drive or drives to be tested as shown in Table 9.2.

Table 9.2. Switch Register Selection of Dectape Unit in the Random Exerciser Test.

Dectape Unit on Thumbwheel Switch	Octal Value of Switch Register
0	4000
1	2000
2	1000
3	0400
4	0200
5	0100
6	0040
7	0020

Several drives may be tested at one time by combining the switch register settings for the corresponding Dectape drives. Press CLEAR and CONTINUE switches. The program will halt again. Place 0000 in the switch register and press CONTINUE. To halt the test, press the HALT switch. The complete program documentation is required to interpret the error messages and use all program options.

Errors will be found if a read/write head is dirty, the tape is badly worn, or the unit is overheating due to an inoperable cooling fan. A worn tape may be reclaimed by reformatting. It is recommended that the first five feet of a worn tape be removed before reformatting. This may eliminate the most worn section. Other errors may be due to an incorrect tape drive motor brake adjustment. This undesirable situation can also be observed by sluggish performance of the tapes on stops and turn arounds. This problem can wipe out existing files on a tape or make it impossible to store sample data.

To correct this problem on a dual transport TU 56 tape unit, there are two adjustments on printed circuit board M302 which is the dual delay motor voltage. The bottom trimpot adjusts the right transport and the top trimpot adjusts the left transport. Whether the pots should be turned clockwise or counterclockwise depends on which motor is most out of adjustment. The adjustment is most easily made if one person operates the transport manually while the second adjusts the trimpot. Adjust voltages until stops and starts are crisp.

DISK DIAGNOSTIC PROGRAM

This extremely important program has several functions and it is described separately for the reasons given in the section on computer diagnostics. The program's functions include formatting new disk cartridges, testing the surface of the magnetic disk, testing the functions of the disk drive, and testing the disk controller. If any errors are found, an error number is printed. The error numbers and explanations of the malfunctions are listed in the System Industries disk maintenance manual. These error numbers and explanations have little or no meaning to the average operator, but it is important to run the diagnostic program and have the error message ready before calling a service engineer.

The disk diagnostic program is named DDIAGN and it may be loaded into the computer from an OS-8 disk as indicated below. In the following discussion, the SWR refers to the switch register on the PDP-8 front panel. The first prompt printed by the disk diagnostic program is:

ENTER DISK TEST INFORMATION

After this and all other prompts, the computer halts (the run light goes out). The procedure followed is the same for all prompts: Enter a twelve digit binary number into the SWR and press the CONTINUE switch on the PDP-8 front panel. In the instructions that follow the twelve digit binary number is expressed as a four digit octal number. The sample dialogue is recommended for most purposes, but some optional responses to several prompts are discussed below. Insert and start the OS-8 disk.

```
.R DDIAGN

ENTER DISK TEST INFORMATION
```

Before responding to this prompt, remove the OS-8 disk and insert a new or test cartridge. Enter 300x into the SWR where x = the disk unit number (0 for a single disk drive, 1 for the second unit of a two drive system, etc.) and press continue. If more than one drive is present, more than one test can be started at the same time as follows:

```
ENTER DISK TEST INFORMATION
(enter 3000 and press continue)
ENTER DISK TEST INFORMATION
(enter 3001 and press continue)
ENTER DISK TEST INFORMATION
(enter 0000 and press continue)
ENTER UNIT ADDRESS
(enter 0050 and press continue)
ENTER FORMAT LOOP CONTROL
(enter 0000 and press continue)
ENABLE FORMAT SWITCH, ERROR AND TYPE OUT
CONTROLS
(enter 0002 and press continue)

UNIT 0
UNIT 1

DISABLE FORMAT SWITCH

UNIT 0
UNIT 1
```

Each time the unit number is printed a new cycle of tests begins for that drive. Note that any and all programs or data stored on the test disk will be destroyed. One of the prompts printed is:
ENTER FORMAT LOOP CONTROL
If a 0000 is entered, the disk will be formatted once at the beginning of the first pass of the tests. On subsequent passes, the disk will not be reformatted. Any non-zero entry will cause the program to reformat the disk before every

pass of the tests. A single formatting is adequate and 0000 should be entered. The last prompt printed is:

ENABLE FORMAT SWITCH, ERROR AND TYPEOUT CONTROLS

Several responses to this prompt are possible to perform a variety of functions. Place the format enable switch on the rear panel of the disk controller in the format position, enter one of the following codes into the switch register, and press CONTINUE. Code 0002 is recommended for most situations and is especially suited to over-night testing for intermittent errors.

0002. Causes the disk to be formatted and runs diagnostics. If an error occurs, an error message is printed and the program continues with the next test. At the conclusion of all tests the sequence of tests is repeated. The program runs about one hour with the format switch enabled and completes one cycle of tests in another hour with the format switch disabled.

0000. Causes the disk to be formatted and runs diagnostics. If an error occurs, an error message is printed and the program halts. Running time is the same as 0002.

0003. This entry is used following a halt with 0000. Enter 0003 and press CONTINUE and the diagnostic will continuously repeat the test that caused the error.

0040. This is a fast format option. The disk is formatted in about two minutes and the program halts without requesting the user to disable the format switch. This option is acceptable in emergency situations when a formatted disk is required immediately. It is not recommended in general as all disks should be thoroughly tested after formatting.

To stop the test, press the HALT switch. The OS-8 disk may be reinserted and the OS-8 system restarted at address 7600. Alternatively a new test disk may be inserted and DDIAGN restarted at 0200.

9.3 SYSTEM ZERO ADJUSTMENT PROBLEMS

If the system zero adjustment described in section 2.4 cannot be achieved, or if the signal is excessively noisy, there are several likely causes. Instabilities are sometimes caused by overheating of electronics due to a cooling fan failure or a dirty filter that blocks the passage of an adequate amount of cool air. Vacuum tubes in the RF generator (1015 and 3000 series) and power supplies should be checked or replaced. The preamplifier may be faulty and zeroing should be attempted on all of the preamp ranges.

If possible swap the preamp for a spare and attempt to zero. Some preamplifiers have a coarse zero screwdriver adjustment. Set the fine zero pot to mid-range and adjust the coarse zero using the ZERO program. Some preamplifiers have a screwdriver hum balance adjustment. Consult the manufacturers manual for details of this adjustment.

The other potential major problem area is the mass spectrometer interface. The proper functioning of the interface is best tested using the interface diagnostic program described in section 9.4. If the diagnostic program indicates the interface is functioning properly, yet it is not possible to zero, the fault may be a noisy electron multiplier. Before replacing or reconditioning the multiplier, refer the subsequent sections including LOSS of IONS. As with many other problems, failure to zero properly also suggests a dirty ion source.

9.4 MASS SCALE CALIBRATION PROBLEMS

The automatic mass scale calibration program should complete a successful calibration of the mass range 20–650 amu at a PFTBA pressure of about 1 X 10^{-6} Torr in about 40 seconds or less. If a longer time is required, or the system will not calibrate, moderate to serious trouble is indicated. There are a very large number of potential causes for mass scale calibration problems and it is simply not possible to describe each in detail. Therefore a general approach to the diagnosis of the problem is given here and the manufacturer's maintenance manual should be consulted for specific details.

After a calibration failure, the system tune-up should be checked following the procedure in the section 2.2. During this tune-up check it may not be possible to observe any signals from PFTBA ions. This is classified as a major problem and one should check the next section titled LOSS OF IONS. If ions are observed the calibration problem should be attacked in the following manner.

If the calibration diagnostic program described in section 2.5 was routinely applied, past records should be examined for signs of day to day drift. The vacuum tubes in the RF generator, particularly the 6AL5 or its 5726 replacement option, are prime suspects in cases of calibration drift. Problems with these tubes may not be observed with a vacuum tube checker. A new tube should be inserted and its stability determined experimentally. If drift continues, try several new tubes and select the most stable one of the group. The Finnigan 3000 series instruments have four vacuum tubes including a 6AL5 in the RF generator.

A drift in an RF tuning capacitor can cause calibration problems. This capacitor compresses or expands the mass range observed during a single sweep of the rod voltages. There is one tuning capacitor for each range on the 1015 and a single tuning capacitor for the 3000 series mass range. The proper adjustment of the tuning capacitors is described in the first mass, last mass calibration procedures in the Finnigan 1015 operator's manual. A similar procedure exists for the 3000 series spectrometers.

A problem in the mass spectrometer interface can cause calibration failures. The proper functioning of the interface diagnostic program is described in a subsequent section of this chapter. Another valuable program for diagnosing instability problems is the peak profile diagnostic program which is also described in this chapter. Again a dirty ion source may be the cause of a calibration failure.

INTERFACE-SPECTROMETER GAIN ALIGNMENT

Calibration problems with the Finnigan 1015, 3000, and 4000 mass spectrometers could be caused by incorrect alignment with the PDP-8 interface. The System Industries (SI) interface outputs 0–10V which corresponds to 1–750 amu or 1–1000 amu. The amu range is handled by the driver file. The standard calibration driver file that is built into the calibration program is set up for 43 digital to analog converter (DAC) units per amu (43 x 750 = 32,250). The 15 bit DAC has a range of 0–32,768 DAC units. The FC1000 and M1000 driver files are set up for 32 DAC units per amu (32 x 1000 = 32,000).

The problem that develops is that the 0–10V range of the SI interfaces have to be matched with the 0–10V range of the Finnigan mass spectrometers. The calibration files work as follows. The driver files consist of a list of DAC ranges that correspond to the calibration masses. The DAC ranges cover ±3 amu around the calibration mass. For example in the standard driver file mass 100 has a DAC range of 43 x 100 = 4300 DAC units ± (3 x 43 =) 129 DAC units; in other words mass 100 ± 3amu corresponds to the range of 4171–4429 DAC units. The SI interface outputs the 4171–4429 DAC unit range which corresponds to 1.2728–1.3516V sent to the spectrometer. The most important idea is that the mass 100 peak must be in this range and all other calibration masses must be in their ranges. On the Finnigan 1015 R-73 on PC5 is an offset adjustment to make mass 100 fall in the correct range.

If the Finnigan ramp voltage (PC-5 on 1015) is set up so that mass 100 is passed through the quadrupole at 1.262 volts, calibration will fail because the SI interface and calibration program look for mass 100 between 1.2728–1.3516V. R-73 (on the 1015) shifts the ramp up or down without changing the slope. Thus the spectrometer may be set up so mass 100 will be passed through the quadrupole in the correct voltage range.

A problem was created when Finnigan eliminated the R-73 adjustment in the 3000 and 4000 series mass spectrometers. R-73 may be thought of as a very coarse resolution, coarser than R-74. To compensate there is a "MASSOUT GAIN" pot in the new RIB interface. This allows matching of the interface ramp with the instrument ramp. To make this adjustment run the peak profile diagnostic program MSSCAN, which is described in section 9.4 and introduce PFTBA. Center mass 100, or 219, or etc. on the screen with about a 500 DAC unit range. Then calculate where the peak should be in DAC units as follows: 1–1000 amu instruments: mass x 32 DAC units = ______; 1–750 amu instruments : mass x 43 DAC units = ______.

Using MSSCAN determine where your peak is, and adjust R-73 on the 1015 or the MASSOUT GAIN on th RIB until the mass is centered about where it should be. Remember there is a ±3 amu (± 96 or 129 DAC unit) range in the calibration file. The user should be aware that the MASSOUT GAIN pot may be mounted over the letters SSOUT; therefore it looks like MA GAIN on the label in the RIB interface.

LOSS OF IONS

If no ions are observed during an attempted tune-up, the first check is the 0–200V ramps. On the model 1015 these are observed at TP5 and TP6 on PC5 (see Figure 2.1). On the 3000 series instruments the ramps are observed at TP4 and TP5 on RF/DC control card. If the ramp is present the loss of ions is probably due to a failure in the ion source, electron multiplier, or one of their power supplies. A non-existent ion source collector (trap) current suggests a broken filament; a lower than normal trap current suggests a sagging filament that must be replaced. If source potentials are missing or low in value, the problem may be on the ion source control PC board. The electron multiplier power supply output should be checked. Some power supplies have vacuum tubes that should be replaced regularly. See chapter 8 for a recommended replacement schedule.

Another good trouble shooting approach is to shut down all power to the mass spectrometer and check all electrical leads to the ionizer and multiplier for continuity or no continuity as appropriate. Consult the manufacturers maintenance manual for specific directions. If shorts are indicated it will be necessary to remove the ionizer and multiplier assemblies from the manifold and check for shorts and correct assembly.

If the –200V ramp was not present, check TP20 on PC5 of the 1015 for the ramp. Check TP1 and TP6 on the RF/DC control board of the 3000 series instruments. If this is present the problem is probably a component on the PC5 board or the RF/DC control board. If the ramp is not present at

TP20 (TP1/6) check the output of the ±1200V and +200V power supply. Vacuum tubes should be replaced if necessary. Also there are vacuum tubes in the RF generator of the 1015 and 3000 series instruments. See chapter 8 for more details on these.

INTERFACE DIAGNOSTIC PROGRAMS

Problems in the mass spectrometer interface are often revealed during an attempt to calibrate or during operations with an apparently acceptable calibration. The interface diagnostic program is an excellent approach to finding interface problems. It is a good idea to run the interface diagnostic program at regular intervals as a preventive maintenance routine when the GC/MS system is not in use. This will exercise the electronic components of the interface and familiarize the operator with the method of running the program.

There are two versions of the program that correspond to two models of the interface. Both models perform identical functions, but the later model takes advantage of advances in solid state technology and is a smaller overall package that may be somewhat remote from the data system. This interface is often called the remote in base (RIB) interface. The interface diagnostic programs are identified with hardware as follows:

Interface Model Number	Interface Diagnostic Program
1152–7016 (separate analog and digital drawers)	MSTEST
5000–7073 (RIB or remote interface)	NEW150

The interface diagnostic programs are stored on the OS-8 system and may be loaded into the computer as described in the previous section on computer diagnostic programs. Both programs have a starting address of 0200. For general trouble shooting surveys, operator familiarization, and interface exercising, the followiing responses to the prompts are strongly recommended:

R MSTEST

```
DO YOU WANT STATIC TEST? NO
DO YOU WANT MASS SPEC TEST?  YES
DO YOU WANT OFFSET ADJUST PROMPT?  NO
DO YOU WANT TRACKING ADJUST PROMPT?  NO
DO YOU WANT IFSS TEST?  YES
```

```
DO YOU WANT PLOTTER TEST?  NO
PUT LOGIC DRAWER IN TEST MODE (place manual test
      switch on back of logic drawer in ON position)
      (press continue)

OK
OK
OK
OK

      (after completion of test, place manual test
      switch in off positon)

.R NEW150

DO YOU WANT STATIC TEST?  NO
DO YOU WANT TRANSMISSION TEST?  NO
DO YOU WANT ANALOG ADJUST?  NO
DO YOU WANT MASS SPEC TEST?  YES
DO YOU WANT REMOTE SEQUENCING TEST?  YES
DO YOU WANT IFSS TEST?  YES
DO YOU WANT PLOTTER TEST?  NO
OK
OK
OK
```

The prompts that are answered in the negative refer to specialized tests and adjustment aids that are best used after trouble is indicated in the Mass Spec and/or IFSS tests. It is extremely important to run the IFSS test since this mode is used during mass scale calibration. A major advantage of running the suggested tests only is that after completion of each test, the program prints OK, then repeats the diagnostic. Therefore it is possible, and strongly recommended, to run the test continuously over a long period of time, including overnight. This will enhance considerably the chances of finding crucial but intermittent errors. Be certain to enter 0000 in the switch register to insure the program will continue even if an error is detected. To stop the program press the HALT switch.

If an error is detected, the program will print an error message. As usual most of the error messages will probably have little meaning to the operator even after consulting the appropriate System Industries mass spectrometer interface maintenance manual. However having the error numbers available when a service engineer is called could save considerable time and expense.

However, there are several errors that are related to relatively simple potentiometer (pot) adjustments that should be attempted by the operator. Errors 3 and 4 occur if the analog to digital converter (ADC) zero is not in agreement with the digital to analog converter (DAC) zero. The zero may be adjusted by a trial and error procedure until errors 3 and 4 are corrected. Trial and error consists of turning the zero adjust potentiometer on the right side of the ADC (Phoenix analog drawer) one-half turn in one or the other direction while the diagnostic is running and observing the errors messages. A YES response to the MSTEST prompt DO YOU WANT OFFSET ADJUST PROMPT? was intended to facilitate this adjustment by continuously rerunning the diagnostic that leads to errors 3 and 4. However there is little point in continuously rerunning the one specific diagnostic, since the entire MASS Spec and IFSS tests only require several minutes. A much quicker method of making this adjustment in the two drawer interface is to run the program ADZERO.

ADZERO Program. This program is a modfied version of the preamplifier zero program, ZERO, described in section 2.4. It is loaded as follows:

```
SELECT MODE:  ADZERO
MANUAL?  N
AUTOMATIC?  N
ZERO ADJUST?  Y
```

This program causes the DAC to output zero volts. This output is then connected (through software) to the input of the ADC with the ADC's output being displayed in the accumulator of the PDP-8. If no lights are visible, turn the zero adjust pot CCW until a light appears in the accumulator. Rotate the pot CW until the light goes out completely (stops flickering). At this point rotate the pot another 1/2 turn CW. This will insure operation in the safe region of the operating range. If a light or lights are initially displayed in the accumulator, eliminate them using the above procedure.

The small remote interface may require the same adjustment after an occurrence of the same errors. The correct pot is labeled integrator offset on the top board of the interface. The ADZERO program is fully compatible with the remote interface and may be used to facilitate the adjustment.

Another error that may be corrected by a simple adjustment on either interface is error 6. This is caused by an improper gain pot adjustment. This pot on the remote interface is labelled the unity gain pot. A trial and error process should be used during the Mass Spec and IFSS tests to make this adjustment.

With the excepton of the Houston plotter test, the other prompts of MSTEST and NEW150 were intended to facilitate certain adjustments and measurements. These are not as generally useful to the system operator, but may be used if desired. The System Industries interface manuals describe these tests. The plotter test may be the most useful of the remaining tests, but a failure does not necessarily indicate trouble in the plotter itself. The plotter interface is contained in the digital bay of the two drawer interface, and is on a separate PDP-8 board when the remote interface is present.

PEAK PROFILE DIAGNOSTIC PROGRAM

This program is a valuable tool to investigate problems of system zero adjustment, calibration, quality assurance, or intermittent instability. For example, if the mass scale calibration is failing because of a weak 6AL5 (5726) detector diode tube in either a model 1015 or 3000 series, a clear observation of this instability can be made with the program.

The program gives the operator the capability of observing repeatedly an entire mass spectrum, a small portion of a mass spectrum, or a single peak in a spectrum at the data system CRT. These same basic capabilities are available at the mass spectrometer console oscilloscope, but the CRT is easier to use for diagnostics because of its storage capability. With the CRT and the program, the operator has a very much better opportunity to examine clearly the details of peak shape and its stability. Also since this program uses the data system, if a spectrum is clearly observed on the oscilloscope, but not with the program, the problem is immediately isolated to the data system itself or the data system-mass spectrometer interface.

The name of the diagnostic program is MSSCAN and the program is available on the OS-8 and the real time operating systems. Before the program is run, perfluorotributylamine (PFTBA) must be introduced into the mass spectrometer. Turn on the ionizer and the electron multilplier. Under OS-8 the program is called with the command R MSSCAN. Under the real time system the dialogue is as follows:

```
SELECT MODE:  MSSCAN
CONST, SCAN OR?
```

A user entry of a question mark causes the program to print a list of MSSCAN commands that is reproduced and explained in detail in this section. A user entry of the command C (not CONST) causes the program to enter a non-scanning mode. In this mode additional commands are available as shown below. One may cause the program, by the various keyboard entries to set and hold various rod voltages. These voltages are expressed in terms of the corresponding digital to analog converter (DAC) units. Also one may print the corresponding analog to digital converter (ADC) data after integration of the electron multiplier signal for various time periods. This constant mode of operation has some utility for trouble shooting, but is is generally less useful than the other mode of operation.

The most useful application of the program is the scanning mode that is entered by the command S (not SCAN). Subsequent prompts are shown in the sample dialogue, which should cause mass 219 (from PFTBA for example) to be displayed near the center of the CRT screen with a 1–750 amu mass range instrument.

```
CONST, SCAN OR ?S
STARTING DAC 9150
RANGE  500
INCREMENT  3
```

If the peak is not prominently displayed, any of the single character commands under the scan mode may be entered to modify the program scanning conditions. The screen will be paged automatically, and the command printed. Enter the new value and scanning will be resumed. The first command to try is the M command which multiplies the peak intensity by the value entered. One may also change the starting DAC, range and increment slightly since the exact values required to display mass 219 in the center of the CRT will vary somewhat from instrument to instrument. In general one amu corresponds to about 43 DAC units on a 750 amu mass range intrument and 32 DAC units on a 1000 amu mass range instrument. A starting DAC of about 2000, a range of 20000, and an increment of 10 will display the spectrum from about 40–700 amu. To exit from the program or to recover from an error after paging the screen manually, enter the E command. This will return the program to the first prompt. Under OS-8 enter a CONTROL/C and under the real time system a CONTROL/L to return to the system monitor.

MSSCAN Variables

C — constant voltage (DAC)

S — set DAC and read ADC
U — increment DAC and read ADC
D — decrement DAC and read ADC

P — print current DAC
T — set time register (max. 4095)
R — repeat integration at current register settings
E — back to start

S — scan on 4010 display

S — change starting DAC
R — change range
I — change increment
T — change time register
P — print parameters
O — one step each time space bar is pressed
C — clear one step - full speed
M — multiply intensities
D — divide intensities
E — back to start

Integration time will be equal to the value in the time register, in milliseconds. This register is set to 1 at the start of the program.

The most important observations to make from the mass 219 ion of PFTBA are peak shape, resolution from the mass 220 ion, position stability, and abundance stability. The peak should be symmetrical without excessive noise or a bra shaped top. Its stability in position and abundance should be very constant with little or no movement. It may be necessary to examine a broader mass range, e.g., 50 amu, to adequately judge position stability. A noisy or unstable condition is a clear sign of trouble and a frequent cause of mass scale calibration failures.

Some of the conditions observed with MSSCAN are listed below together with a few possible causes of the problem.

Position Instability: A bad 6AL5 (5726) tube in the model 1015 or 3000 series RF generator; also three 807 RF tubes are contained in the 1015 generator and two 8458 tubes and a 6EB8 are contained in the generator of the 3000 series instruments. The power supply for the model 1015 has vacuum tubes that can cause severe instability.

Abundance Instability: A bad preamp; bad tubes in the electron multiplier power supply or a bad electron multiplier; a sagging filament or unstable ion source potentials.

Noisy Peaks: Improper grounding which produces 60 Hz noise in a ground loop; bad power supply tubes or unstable source voltages. The mass set voltage at TP18 on PC-4 of theFinnigan 1015 should be stable to better than a millivolt regardless of the source, i.e., internal or datasystem.

9.5 QUALITY CONTROL PROBLEMS

Quality control problems are defined as problems that are recognized only when examining the overall peak size caused by a known amount of DFTPP or by examining the mass spectrum of DFTPP. At this stage the system should be tuned and the mass scale calibrated. The quality control performance evaluation is conducted as described in section 2.6. Examine the output from the evaluation and consult the appropriate section that follows.

POOR OVERALL SENSITIVITY

Poor overall sensitivity may be caused by a variety of factors and is roughly judged by the size of the peak obtained from 20ng of DFTPP. The operator must be familiar with the capabilities of the instrument when it is in good working order. The following list of possible causes of poor overall sensitivity is ordered from the first things to check to some actions of nearly last resort:

1. For packed columns connected to a jet separator, the mass spectrometer pressure should be about 10^{-5} mm of Hg. A lower pressure suggests a plugged separator; check the temperatures of the separator and transfer line oven for a possible heater failure. A cool oven will reduce sensitivity by condensation of the test compound.
2. The dilute standard solution of DFTPP should be remade if it is not fresh. Old solutions of all dilute standards suffer concentration losses due to adsorption of trace quantities on the walls of containers or oxidation by trace quantities of oxygen or other oxidizing agents.
3. If the shape of the peak is broader than normal or exhibits tailing, the column may be poorly conditioned or suffering from old age. Also the metal fritted filter or separator may be partly plugged.
4. The tune-up should be checked for over resolved peaks in any or all mass ranges. Use only as much resolution as needed as illustrated in section 2.2. Note carefully the proper functioning of all tune-up potentiometers. Recalibrate and repeat the DFTPP performance evaluation.
5. The peak profile diagnostic program should be run with particular attention to noisy peaks and possible ground loop problems and other instabilities as discussed in section 9.4.
6. Clean the ion source, check for a sagging filament, and clean the rods according to the manufacturers's procedures.

7. If all else fails, recondition electron multiplier using procedures from the manufacturer's manual.

POOR HIGH MASS SENSITIVITY

This condition is revealed by unacceptable ion abundance at masses 365 and 442 in the DFTPP spectrum. Poor high mass sensitivity is most frequently caused by dirty rods or over resolution of masses 219 and 220 and 502 and 503 as in manufacturer's tune-up procedure. If a tune-up according to the EPA procedure in section 2.2 is not successful, clean the ion source and rods.

INCORRECT MASS ASSIGNMENTS OR POOR RESOLUTION OF ADJACENT IONS

This condition is revealed by an examination of ion abundances in the DFTPP spectrum at masses 68–70, 197–199 and 441–443. If the mass scale calibrated correctly, but incorrect mass assignments appear in the DFTPP spectrum, a slow drift of the mass scale calibration is indicated. This is most frequently caused by a weak 6AL5 tube in the RF generator (1015 and 3000 series instruments). Refer to section 9.4 and locate the cause of the drift.

If the mass assignments are correct but the ions at masses 68, 197, or 441 are too abundant, front end lift off is indicated. This condition is often caused by an incorrect alignment of the source magnet or incorrect adjustment of the ion source potentials (lens, extractor, ion volume, etc.).

Insufficient resolution will cause excessive ion abundance at masses 68, 197 or 441 as well as at masses 70, 199, or 443. Acceptable ion abundance at masses 68, 197, or 441, but inaccurate 69/70, 198/199, or 442/443 ratios may be caused by a system zero adjustment that sets the threshold too high and clips off real ion abundance data.

HIGH BACKGROUND

High background is defined as the presence of extraneous ions in an otherwise acceptable spectrum of 20ng of DFTPP. A more or less continuous band of ions across a broad mass range suggests a poor system zero adjustment, in which the threshold is set too low. Repeat the zero adjustment in section 2.4 and make sure one accumulator light is either just flashing or just off. If a steady zero cannot be achieved, see section 9.3 for possible causes.

Chemical background can result from a serious accumulation of pump oil in the manifold, solid probe, or transfer line flange area. This could be caused by a lengthy power failure and the resultant diffusion of oil vapor or by a failure in the overheat protect circuit that allowed a diffusion pump to overheat. Removal of pump oil is accomplished by a thorough internal cleaning of the system with solvents and the manufacturer should be consulted for assistance.

Chemical background can also result from liquid phase bleeding from the GC column, contaminated carrier gas, a contaminated injection system, or a small leak in the GC column. As appropriate the column should be operated at a lower temperature, reconditioned or changed. If this fails to eliminate the background change or regenerate the carrier gas filter drying trap or change the carrier gas cylinder. Injector contamination can be eliminated by a thorough cleaning of the injection port with a series of solvents. However, if on-column injection is employed, cleaning the injection port will not help. If the head of the column is highly contaminated, discard the old packing, thoroughly clean the column, and repack.

CHAPTER 10
SELECTED BIBLIOGRAPHY

It is not the purpose of this chapter to present a comprehensive listing of all the known literature of GC/MS and related articles on enviriomental applications. Rather a limited number of books, review articles, and original journal articles are cited which are considered important contributions and keys to further information about these subjects.

10.1 GENERAL REFERENCE BOOKS

S. Safe and O. Hutzinger, "Mass Spectrometry of Pesticides and Pollutants", CRC Press, Cleveland, OH, 1973.

R.E. Gould, "Fate of Organic Pesticides in the Aquatic Environment", Advances in Chemistry Series No. 111, American Chemical Society, Washington, D.C., 1972.

F.W. McLafferty, "Interpretation of Mass Spectra", 2nd ed., W.A. Benjamin, Inc., New York, N.Y., 1973.

G.W.A. Milne, "Mass Spectrometry, Techniques and Applications", Interscience Publishers, John Wiley and Sons, Inc., New York, N.Y., 1971.

H. Budzikiewicz, C. Djerassi, and D.H. Williams, "Mass Spectrometry of Organic Compounds", Holden-Day Inc., San Francisco, CA, 1967.

J. Roboz, "Introduction to Mass Spectrometry, Instrumentation and Techniques", Interscience Publishers, John Wiley and Sons, Inc., New York, N.Y. 1968.

W. McFadden, "Techniques of Combined Gas Chromatography/Mass Spectrometry-Applications in Organic Analysis", Interscience Publishers, John Wiley and Sons, Inc., New York, N.Y.,1973.

G.R. Waller, "Biochemical Applications of Mass Spectrometry", Interscience Publishers, John Wiley and Sons, Inc., New York, N.Y., 1971.

Q.N. Porter and J. Baldas, "Mass Spectrometry of Heterocyclic Compounds", Interscience Publishers, John Wiley and Sons, Inc., New York, N.Y., 1971.

F.W. McLafferty, "Mass Spectrometry of Organic Ions", Academic Press, Inc., New York, N.Y., 1963.

J.H. Beynon, R.A. Saunders, and A.E. Williams, "The Mass Spectra of Organic Molecules", Elsevier Publishing Co., New York, N.Y., 1968.

K. Biemann, "Mass Spectrometry-Organic Chemical Applications", McGraw-Hill Book Co., Inc., New York, N.Y., 1962.

M.C. Hamming and N.G. Foster, "Interpretation of Mass Spectra of Organic Compounds", Academic Press, New York, N.Y., 1972.

H.M. McNair and E.J. Bonelli, "Basic Gas Chromatography", Varian Aerograph, Walnut Creek, CA, 1969.

W.R. Supina, "The Packed Column in Gas Chromatography", Supelco, Inc., Bellefonte, PA, 1974.

T.R. Lynn, "Guide to Stationary Phases for Gas Chromatography", Analabs, Inc., North Haven, CT, 1975.

B.J. Millard, "Quantitative Mass Spectrometry", Heyden & Son Inc., Philadelphia, PA, 1977.

10.2 PRINTED MASS SPECTRA COLLECTIONS

A. Cornu and R. Massot, "Compilation of Mass Spectral Data", 2nd ed., Heyden and Son, Ltd., London, England, 1975.

"Eight Peak Index of Mass Spectra", Vol. 1–3, Mass Spectrometry Data Centre, Atomic Weapons Research Establishment, Aldermaston, England, 1975

E. Stenhagen, S. Abrahamssen, and F.W. McLafferty, "Registry of Mass Spectral Data", Interscience Publishers, John Wiley and Sons, Inc, New York, NY, 1974.

"EPA/NIH Mass Spectral Data Base", National Bureau of Standards, National Standard Reference Data Service publication No. 63 (1978).

10.3 SELECTED REVIEW AND PRIMARY JOURNAL ARTICLES

MASS SPECTROMETRY

Burlingame, A.L., Shackleton, C.H.L., Howe, I., Chizhov, O.S., Anal. Chem. Annual Reviews, 50, 346R (1978).

Burlingame, A.L., Kimble, B.J., and Derrick, P.J., Anal. Chem. Annual Reviews, 48, 368R (1976).

Burlingame, A.L., Cox, R.E., and Derrick, P.J., Anal. Chem. Annual Reviews, 46, 248R (1974).

Burlingame, A.L. and Johanson, G.A., Anal. Chem. Annual Reviews, 44, 337R (1972).

DeJongh, D.C., Anal. Chem. Annual Reviews, 42, 169R (1970).

Large, R. and Knof, H., Org. Mass Spectrom., 11, 582 (1975).

Fales, H.M., Milne, G.W.A., Winkler, H.V., Beckey, H.D., Damico, J.N., and Barron, R., Anal. Chem., 47, 207 (1975).

Alford, A., Biomed. Mass Spectrom., 2, 229 (1975).

Heller, S.R., Milne, G.W.A., and Feldmann, R.J., Science, 195, 253 (1977).

GC/MS

Fenselau, C., Appl. Spectrosc., 28 , 305 (1974).

Brooks, C.J. and Middleditch, B.S., Mass Spectrom., 2, 302 (1973).

Junk, G.A., Int. J. Mass Spectrom. Ion Phys., 8, 1 (1972).

Oswald, E.O., Albro, P.W., and McKinnery, J.P., J. Chromatogr., 98, 363 (1974).

Heller, S.R., McGuire, J.M., and Budde, W.L., Environ. Sci. Technol., 9, 210 (1975).

Eichelberger, J.W., Harris, L.E., and Budde, W.L., Anal. Chem., 47, 995 (1975).

Gates, S.C., Smisko, M.J., Ashendel, C.L., Young, N.D., Holland, J.F., and Sweeley, C.C., Anal. Chem., 50,433 (1978).

DIRECT AQUEOUS INJECTION

Harris, L.E., Budde, W.L., and Eichelberger, J.W., Anal. Chem., 46, 1912 (1974).

"Standard Methods for the Examination of Water and Wastewater", 13th Edition, 1971.

Baker, R.A., J. Amer. Water Works Ass., 58, 751 (1966).

Fujii, T., J. Chromatogr., 139, 297 (1977).

INERT GAS PURGING AND TRAPPING

Bellar, T.A. and Lichtenberg, J.J., J. Amer. Water Works Ass., 66, 739 (1974).

F.C. Kopfler, R.G. Melton, R.D. Lingg, and W.E. Coleman in "Identification and Analysis of Organic Pollutants in Water", 1st ed., L.H. Kieth, Ed., Ann Arbor Science Publishers Inc., Ann Arbor, MI, 1976. Chapter 6.

Bellar, T.A., Lichtenberg, J.J., and Kroner, R.C., J. Amer. Water Works Ass., 66, 703 (1974).

"The Analysis of Trihalomethanes in Finished Waters by the Purge and Trap Method", Environmental Monitoring and Support Laboratory, U.S. Environmental Protection Agency, Cincinnati, OH, 1977.

Bellar, T.A. and Lichtenberg, J.J., "Semi-Automated Headspace Analysis of Drinking Waters and Industrial Waters for Purgeable Volatile Organic Compounds" submitted for publication to ASTM, July 1978.

Bellar, T.A., Lichtenberg, J.J. and Eichelberger, J.W., Environ. Sci. Technol., 10, 926 (1976).

Coleman, W.E., Lingg, R.D., Melton, R.G. and Kopfler, F.C., in "Identification and Analysis of Organic Pollutants in Water", L.W. Keith, Ed., 305–327, Ann Arbor Science, Ann Arbor, Mich., 1976.

Symons, J.M., Bellar, T.A., Carswell, J.K., DeMarco, J., Kropp, K.L., Robeck, G.G., Seeger, D.R., Slocum, C.J., Smith, B.L. and Stevens, A.A., J. Amer. Water Works Assoc., 67, 634, (1975).

Brass, H.J., Feige, M.A., Halloran, T., Mello, J.W., Munch, D. and Thomas, R.F., in "Drinking Water Quality Enhancement Through Source Protection", R.B. Pojasek, Ed., 393–416, Ann Arbor Science, Ann Arbor, Mich., 1977.

Bellar, T.A. and Lichtenberg, J.J., presented at the ASTM Symposium on "The Measurement of Organic Pollutants in Water and Wastewater", Denver, CO., June, 1978.

Bellar, T.A., Budde, W.L. and Eichelberger, J.W. in "Monitoring Toxic Substances", D. Schuetzle, Ed., American Chemical Society Symposium Series No. 94, American Chemical Society, Washington, D.C., 1979.

QUALITATIVE HEADSPACE ANALYSIS

McAullife, C., Chem. Tech., 46 (1971).

ADSORPTION WITH POROUS POLYMERS

Junk, G.A., Richard, J.J., Grieser, M.D., Witiak, D., Witiak, J.L., Arguello, M.D., Vick, R., Svec, H.J., Fritz, J.S., and Calder, G.V., J. Chromatogr., 99, 745 (1974).

R.G.Webb, EPA Report EPA-660/4–75–003, Athens, GA, June 1975.

G.R. Harvey, EPA Report EPA-R2–73–177, March, 1973.

Van Rossum, P., and Webb, R.G., J. Chromatogr., 150, 381 (1978).

AIR SAMPLES

Grob, K., J. Chromatogr., 62, 1 (1971).

Lao, R.C., Thomas, R.S., Oja, H., and Dubois, L., Anal. Chem., 45, 908 (1973).

Pellizzari, E.D., Bunch, J.E., Berkley, R.E., and McRae, J., Anal. Chem., 48, 803 (1976).

Bunn, W.W., Deane, E.R., Klein, D.W., and Kleopfer, R.D., Water, Air, and Soil Pollution, 4, 367 (1975).

FATTY TISSUE SAMPLES

Kuehl, D.W. and Leonard, E.N., Anal. Chem. 50, 182 (1978).

CHEMICAL DERIVATIZATION

A.E. Pierce, "Silylation of Organic Compounds", Pierce Chemical Co., Rockford, IL, 1968.

K. Blau and G. King , "The Handbook of Derivatives for Chromatography", Heyden and Son, Ltd., New York, NY, 1977.

Quilliam, M.A. and Westmore, J. B., Anal.Chem. 50, 59 (1978).

SELECTED ION MONITORING

Budde, W.L. and Eichelberger, J.W., J. Chromatogr., 134, 147 (1977).

Eichelberger, J.W., Harris, L.E., and Budde, W.L., Anal. Chem., 46, 227 (1974).

OPEN TUBULAR CLOUMNS

Grob, K., Anal. Chem., 45, 1788 (1973).

Grob, K. and Grob K. Jr., J. Chromatogr., 94, 53 (1974).

Novotny, M., Anal. Chem., 50, 16A (1978).

CHEMICAL IONIZATION

Jelus, B.L. and Munson, B., Biomed. Mass Spectrom., 1, 96 (1974).

Hatch, F. and Munson, B., Anal. Chem. 49, 169 (1977).

Field, F.H., Advan. Mass Spectrom., 4, 645 (1968).

Munson, B., Anal. Chem., 49, 772A (1977).

Hunt, D.F., Stafford, G.C., Crow, F.W., and Russell, J.W., Anal. Chem., 48 , 2098 (1976).

Price, P., Martinsen, D.P., Upham, R.A., Swofford, H.S., and Buttrill, S.E., Anal. Chem., 47 , 190 (1975).

Murata, T., Takahashi, T., and Takeda, T., Anal. Chem. 47 , 573 (1975).

Oswald, E.O., Levy, L., Corbett, B.J. and Walker, M.P., J. Chromatogr., 93, 63 (1974).

Fales, H.M., Milne, G.W.A., and Nicholson, R.S., Anal. Chem., 43 , 1785 (1971).

Holmstead, R.L. and Casida, J.E., J. Ass. Offic. Anal. Chem., 57 , 1050 (1974).

Biros, F.J., Dougherty, R.C., and Dalton, J., Org. Mass Spectrom., 6, 1161(1972).

INDEX